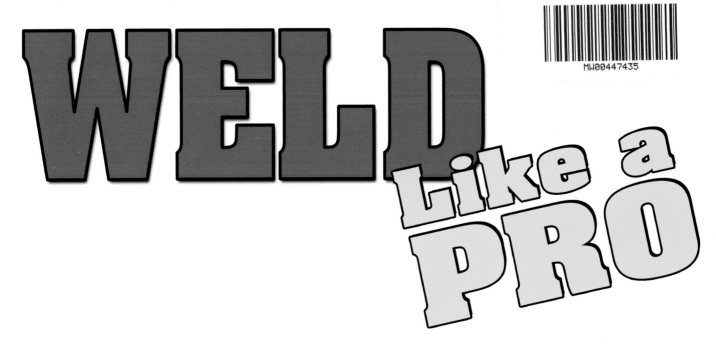

WELD Like a PRO

Jerry Uttrachi

S-A DESIGN

J. SARGEANT REYNOLDS C.C.

CarTech®

CarTech®

CarTech®, Inc.
838 Lake Street South
Forest Lake, MN 55025
Phone: 651-277-1200 or 800-551-4754
Fax: 651-277-1203
www.cartechbooks.com

Edit by Paul Johnson
Layout by Monica Seiberlich

ISBN 978-1-61325-221-5
Item No. SA343

Library of Congress Cataloging-in-Publication Data
Uttrachi, Jerry, author.
Weld like a pro / by Jerry Uttrachi.
pages cm
ISBN 978-1-61325-221-5
1. Welding. 2. Motor vehicles–Maintenance and repair. I. Title.

TS227.U87 2015
671.5›2–dc23

2014048706

Written, edited, and designed in the U.S.A.
Printed in China
10 9 8 7 6 5 4 3

Front Cover Photo: *Courtesy Pat Stanton, Year One.*

Title Page: *Here, a weldor demonstrates stick welding. Developed in the early 1900s, stick welding is very versatile and is widely used in industry. A major advantage of stick welding is that it generates its own shield to protect the weld deposit from atmospheric contamination. It can be used in wind up to about 20 mph.*

Back Cover Photos

Top: *Repairing cast iron requires a specific approach and technique. Cracks should be opened at a sufficiently wide angle to keep the amount of melted cast iron in the weld deposit to less than 30 percent. After preheating the cast manifold, welding progresses in this simulated crack repair. Preheating helps reduce the stresses as the weld cools and is particularly useful for the cast-iron heat-affected zone, which is crack sensitive.*

Middle: *Sheet-metal fabrication projects, such as this inner wheel well and shock tower, require careful spot welding before the seams are filled in with the weld beads. Unibody drift cars require a very rigid chassis so they can respond instantaneously to driver input. Hundreds of spot welds make a single rigid assembly from the many internal body parts. (Jim Harvey Photo, Courtesy Harvey Racing Engines)*

Bottom Left: *This corner TIG weld is made of 1/8-inch-thick magnesium at 130 amps AC with a 65-percent DCSP using argon shielding gas. The DCSP/DCRP balance was altered from the test welds to see if additional cleaning action could be achieved and to reduce the chance of burn-through. The cleaning action was about the same. This autogenous weld (meaning no filler added) wet very well and is porosity free.*

Bottom Right: *This yard gate latch was MIG welded. It required mounting holes that were wider than the 1.25-inch gate frame so it could not be attached as designed. The solution was to narrow the mounting and weld the latch to the gate frame. To make it easier to bring the MIG welder to the fence and because it was welded outdoors, self-shielded wire was used.*

OVERSEAS DISTRIBUTION BY:

PGUK
63 Hatton Garden
London EC1N 8LE, England
Phone: 020 7061 1980 • Fax: 020 7242 3725
www.pguk.co.uk

Renniks Publications Ltd.
3/37-39 Green Street
Banksmeadow, NSW 2109, Australia
Phone: 2 9695 7055 • Fax: 2 9695 7355
www.renniks.com

Canada
Login Canada
300 Ssaulteaux Crescent
Winnipeg, MB, R3J-3T2 Canada
Phone: 800 665 1148 • Fax: 800 665 0103
www.lb.ca

CONTENTS

DEDICATION

To my wife Christine, who tolerated the time required and supported the writing of this book.

ACKNOWLEDGMENTS

I owe a great deal to former colleague and friend Bob Bitzky, who was manager of welding training for ESAB. Bob provided invaluable information and made a number of welds to help demonstrate the benefits of modern welding and cutting systems.

INTRODUCTION

This book is intended for those with an interest in welding or the automotive hobby who may have some welding skills or want to acquire them. It covers modern welding equipment and procedures such as pulsed arc MIG (metal inert gas) and pulsed arc TIG (tungsten inert gas) welding. It will also be of value to anyone who has purchased a MIG welder and wants to understand how its performance and capability compare with other welding processes. It will be useful for someone considering welding as a profession because it covers the basic welding processes used in industry.

This book presents advanced welding topics for fabricating street rods and race cars as well as making a number of common repairs. It covers the welding of carbon steel, chrome-moly steel, aluminum, and stainless steel. An overview of TIG welding titanium and magnesium is included. There are suggestions regarding the proper filler metal choices, why they are selected, and welding techniques to use. Each welding process section includes automotive projects and applications that relate to the process. Details of equipment features are also discussed.

Information for All Skill Levels

This book is not just for the skilled weldor, but an emphasis is placed on advanced techniques for MIG, TIG, oxyacetylene, and stick. With this book, you can learn how to weld various joints, advanced techniques, and processes. It is an excellent source for beginners who want to learn welding and have their work look and perform like a professional's. Some skilled weldors don't believe they need any more than their manual ability. Exceptional manual skills are great to have, and some weldors are on par with the best artists. However, understanding some of the reasons for certain weld problems and why defects occur is also needed. Some welding science is covered, which helps you resolve welding problems.

Modern Arc Welders

Recent advances in welding machines make it easier to produce quality welds. Welding is an art as well as a science and requires skill. However, depositing a stack-of-dimes weld is much easier with the new microprocessor-inverter–based TIG welders; you just preset two current levels: low and high. After the welding current has been on the high setting for a few seconds, filler metal is easier to add. When the current switches to the low setting, moving the torch forward avoids burn-through. Then you set the pulse rate and it switches between the two levels automatically. The current rise and fall times are very quick, providing a very stable arc. When TIG welding with AC power, there is no longer a need for continuous high frequency.

By simply setting the correct plate thickness, the microprocessors in MIG welders automatically set all of the proper parameters. For MIG welding

WELDING GOES MODERN

This book about advanced welding is not just for the skilled weldor. Understanding why welds behave as they do, and some welding science, assists in solving welding problems. Recent advances in welding machines make it easier to produce quality welds. Setting the welding machine to achieve optimum performance has been made much easier with the new microprocessor-inverter-based TIG and MIG welders. (Figure adapted from ESAB's Oxyacetylene Handbook with sketch by Walter Hood)

WELD LIKE A PRO

automotive sheet metal, the short-circuiting mode is used to avoid excess heat. In the past, you had to manually adjust voltage, slope, and inductance to optimize MIG short-arc weld performance; for those who were able to set them properly and possessed good manual skills a quality weld resulted.

Today, the microprocessor monitors the arc and sets the proper electrical characteristics to achieve the optimum short-circuiting conditions. Details of equipment features are presented in the book.

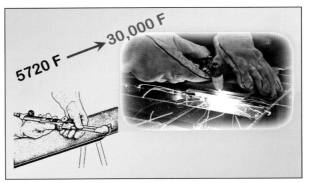

Oxyacetylene cutting produces a very hot flame temperature of 5,720 degrees F. It can make quality cuts in steel. However, plasma cutting severs any metal by using an air stream heated to more than 30,000 degrees F that is forced to flow through a very small hole. The arc intensity is so high that it melts metal rapidly and cuts thin sections of steel much faster than oxyfuel. (Figure adapted from ESAB's Oxyacetylene Handbook *with sketch by Walter Hood)*

Modern Cutting Process

Oxyacetylene cutting produces a very hot flame temperature of 5,720 degrees F. A pure stream of oxygen oxidizes the iron, which generates heat to maintain the cutting process. This produces quality cuts in steel, but it relies on the oxidation of iron to maintain the cut, and therefore it cannot be used on other materials, such as stainless steel and aluminum. However, plasma cutting forces a 30,000 degree F arc and heated air stream through a very small hole to cut through the thickest and strongest material. The arc intensity is so high it quickly melts the metal and cuts thin steel sections much faster than oxyfuel. In fact, plasma cutting melts and severs any metal.

Welding Safety

Welding is safe if you take proper precautions and follow published instructions for the use of equipment and filler metals. The following is an overview of some important safety issues and where to find more safety information.

Material Safety Data Sheets
Before welding, read the safety information from the welding equipment manufacturer. This information includes, but is not limited to, warning labels on the packaging of the welding rods or wires being used and Material Safety Data Sheets (MSDSs) specific to each product. An MSDS lists the potentially hazardous ingredients in a product and provides instructions on how to avoid being overexposed to such ingredients. If an MSDS is not included with your purchase of a welding product, contact your local distributor or the manufacturer of the product directly and ask for one. Most of the MSDS sheets are available on the manufacturers' Web site, free of charge.

Fumes
Because welding fumes can be hazardous, keep your head out of the fumes and always use adequate ventilation. You need to minimize your exposure to welding fumes. Avoid rising smoke and keep any smoke from entering the breathing area under the welding helmet.

Never weld in an enclosed area, such as a car trunk, without some form of forced ventilation, such as a fan. Shielding gases, including both argon and carbon dioxide, are heavier than air and sink to the bottom of an enclosed area, causing a reduction in oxygen leading to dizziness and unconsciousness without warning signs.

Protective Clothing
Aways wear the proper protective clothing. Ultraviolet radiation originating from the arc is harmful to the eyes and skin. Therefore, wear a longsleeved shirt and button it to the top, which blocks arc rays from reaching exposed skin. Special gloves made from thin, flexible leather are available for TIG welding.

You only have two eyes and need to protect them, so always wear safety glasses. Also, use a face shield when grinding. Many eye injuries result from metal particles accumulating in the eyebrows during grinding and falling into the eyes at a later time. On television how-to and restoration shows, weldors do not typically wear long-sleeved shirts and proper clothing, which is an unsafe practice that doesn't set a good example. In fact, such programs generally include a disclaimer indicating that simulations were made for TV viewing and normal safeguards were not used. Unsafe practices seen in those programs should not be imitated. Proper clothing is necessary.

Read and under-stand published safety information for your welding equipment and filler metals. Weld-ing fumes can be hazardous; keep your head out of the fumes and always use adequate ventilation. Exposure to welding fumes can be substantially reduced by avoiding rising smoke and keep-ing the smoke from entering the breathing area under the welding helmet. Never weld in an enclosed area such as a car trunk without some form of forced ventilation.

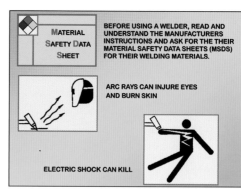

You can find safety informa-tion warning labels on the welding rod and wire packaging and a Material Safety Data Sheet (MSDS) specific to each product. A MSDS lists the potential hazardous ingredi-ents and instructions on how to avoid being overexposed. Contact your local distributor or the manufacturer of the product directly and ask for the MSDS. They are usually available on the manufacturer's Web site.

AWS Literature

For additional details about weld-ing and cutting safety issues, consult ANSI Z49.1, *Safety in Welding, Cut-ting and Allied Processes,* a document available from the American Weld-ing Society (AWS). At the time of this publication, this was available as a free Web site download. In addition, most manufacturers of welding prod-ucts have safety information avail-able on their Web sites.

American Welding Society

AWS is a source for much more than welding and cutting safety information. This nonprofit organi-zation has about 65,000 members and it welcomes anyone to join.

AWS volunteers, who know their respective industries, prepare and write welding procedures and codes for automotive as well as many other areas from airplanes to bridges. Fab-ricators, equipment and filler met-als manufacturers, and government agencies including the military, are members of the volunteer commit-tees. Thousands contribute to these standards and certification programs that are used worldwide. AWS is head-quartered near Miami, Florida, where fewer than 125 people provide the support to the volunteer committees.

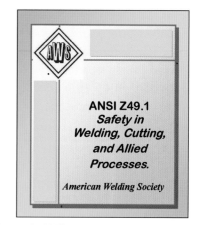

Consult ANSI Z49.1, available from AWS, for additional details about welding and cutting safety issues. At the time of this writing, a free download was available from that organization. In addition, most manufacturers of welding machines, filler metals, and welding gases have safety information available on their Web sites.

The ultraviolet radiation from the arc is harmful to eyes and skin. Always wear a long-sleeves shirt and button it to the top so no arc rays reach any exposed skin. Wear welding gloves with cuffs and always wear safety glasses. You only have two eyes, so protect them. When grinding, use a face shield. Electrical shocks can kill, so don't touch live electrical parts.

AWS is a source for more than welding and cutting safety information. This nonprofit organization has about 65,000 members and anyone is welcome to join. It prepares and publishes welding procedures and codes for automotive as well as many other areas from airplanes to bridges. Volunteers, who know the industry, produce this information. Thousands con-tribute to these standards, which are used and respected worldwide.

WELDING PROCESSES AND EQUIPMENT

This chapter reviews the basic equipment used with each of the welding and cutting processes that are presented in detail in subsequent chapters. Process basics provide an understanding of why and how they work. Detailed equipment specifics of each process are covered in the separate chapters.

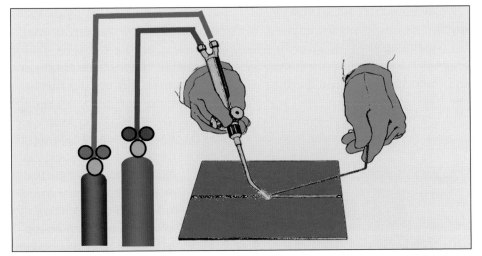

Fig. 1.1. Oxyacetylene welding uses oxygen and acetylene gas supplied in cylinders. Regulators reduce the cylinder pressure and hoses deliver the gases to a torch where they are mixed and exit through a small hole in a welding tip. At the hottest part of the flame tip, the gas mixture burns at 5,720 degrees F. Acetylene is the only fuel gas that can achieve this high temperature in a small concentrated area. (Figure adapted from ESAB's Oxyacetylene Handbook with sketch by Walter Hood)

Oxyacetylene Welding

Oxyacetylene welding is more than 100 years old and is one of the oldest of the welding processes. My old friend, Butch Sosnin, was a welding training consultant and the 1979 president of the American Welding Society. To start all his weldor training, Butch used oxyacetylene because it was slow and students could easily learn welding fundamentals. A student could actually see the molten puddle develop.

After teaching oxyacetylene welding, he followed with TIG welding, which was similar to oxyacetylene in that the heat and filler additions were independent, and there was time to watch the weld puddle develop.

He proceeded to teach stick and MIG welding after some proficiency in the other two methods was achieved. Both stick and MIG welding are more difficult for a beginner to watch because the puddle process happens so fast!

The sequence of this book follows Butch's course instruction and presents the welding processes in the order Butch would teach them.

My friend and colleague, Bob Bitzky, former training manager for ESAB Welding and Cutting Products, agrees with Butch's instructional approach, and starts his new trainees with oxyacetylene followed by TIG welding.

The basic oxyacetylene welding process starts with two cylinders

Fig. 1.2. My old friend Butch Sosnin was a weldor training consultant and the 1979 President of the American Welding Society. Butch insisted on starting his weldor training with the oxyacetylene process because it was relatively slow and students could see the puddle develop. Bob Bitzky, former training manager for ESAB Welding and Cutting Products, supports Butch's logic, stating, "It still holds true today."

of gas—one oxygen and the other acetylene (Figure 1.1). Regulators reduce the cylinder pressure and hoses bring the gases to a torch where they are mixed and exit through a small hole in a welding torch tip. This mixed gas burns at 5,720 degrees F at the hottest part of the flame tip. Other fuel gases may even generate more total heat, but do not have this concentrated, high temperature at the flame tip.

The oxyacetylene flame is the only one that can truly be used for welding. Other fuel gases can be used for cutting and brazing but are not effective for welding. It is the concentration of heat that allows welding to occur. Don't be fooled by just temper-

ature comparison with other gases.

Discussing combustion intensity is a way to explain the temperature concentration of various gases. Without going into too many technical details or specifying the units, an oxyacetylene flame produces more than 10,000 while the next best fuel gas produces about 5,000, or half the value. Steel has the highest melting point of materials that are typically welded. It melts at about 2,500 degrees F. The high-heat concentration and 5,720-degree F flame temperature can melt and fuse two pieces of steel.

Although difficult to master, oxyacetylene welding can be used to weld aluminum. However, unlike steel that turns red then white before forming a molten puddle, aluminum does not. Aluminum melts at 1,200 degrees F and gives little indication it is about to melt.

Note the AWS designation for oxyacetylene welding is straightforward—OAW, although few folks use it.

Gas flow rates for oxyfuel welding are relatively high compared to the shielding gas flow rates in TIG and MIG welding. Needle valves in the torch handle adjust the flow of the two gases. Setting the correct mixture is covered in the individual process section. However, carefully

read the manufacturer's instructions because controlling the flow rate of these gases is very important and potentially a significant safety issue. Be sure to follow the manufacturer's recommendations for adjusting the cylinder regulators. Particularly for the oxygen cylinder, where the pressure adjusting screw must always be backed out before opening the contents valve on the cylinder. Failure to do this properly can cause a surge of high-pressure oxygen to rush into the small chambers of the regulator. Like a Diesel engine, this rapid rise in pressure creates heat and can ignite whatever is in the regulator, including the brass body. In pure oxygen, everything burns, and burns explosively! Follow the manufacturer's recommendations carefully.

TIG Welding

Tungsten inert gas (TIG) welding was first developed in the 1940s to weld aluminum and magnesium. Today it is used to weld many materials, including a variety of steels. TIG creates an arc between a non-consumable tungsten electrode and the workpiece and uses an inert gas, usually argon, to shield the molten puddle. A DC or AC power source supplies the electric power. The tungsten

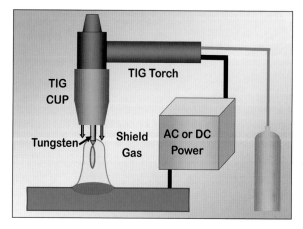

Fig. 1.3. TIG welding uses an inert gas, usually argon, to shield the molten puddle, which is created by an arc between a non-consumable tungsten electrode and the workpiece. Electric power is supplied by a DC or AC source. A major advantage of the process is the ability to have a stable arc at settings as low as low 3 amps.

is held in place with a collet inside a TIG torch. The argon shielding gas protects the molten puddle as well as the tungsten, which may be 6,000 degrees F at the tip. The arc itself is much hotter than the oxyacetylene flame. It is 12,000 to 15,000 degrees F on the outer part of the arc to twice that near the tip of the tungsten. A major advantage of the process is the ability to have a stable arc exist from very low currents (3 amps) up to maximum torch capacity, which can be over 500 amps on automatic machine torches.

For gas tungsten arc welding, the official AWS designation for TIG welding is GTAW. The common term TIG for tungsten inert gas is used throughout the book. Note: If taking welding courses and the instructor insists on the use of GTAW, use it!

Stick Welding

Oscar Kjellberg, the founder of ESAB, invented and was issued a patent for stick electrode welding in 1904. Its use grew rapidly through the 1920s, 1930s, and 1940s and became the leading welding method until displaced by MIG welding.

Stick electrode welding is simple to use, requiring only a power source and an inexpensive electrode holder.

It is still preferred when welding outdoors because it can make acceptable welds in significantly more wind than gas shielding processes. The AWS *Bridge Welding Code,* for example, specifies maximum wind speeds of 5 mph for MIG and TIG welding, while stick welding is allowed up to 20 mph. However, when TIG welding, a gas lens should be used for up to a 4-mph wind. In addition, TIG is more sensitive to the need for quality shielding than MIG.

Stick welding utilizes a simple power source, such as an AC welding transformer. DC power is also widely used and for welding outdoors; portable engine-driven DC generators are often employed. All these power sources are called constant current, referring to setting the desired welding current level, which stays relatively fixed regardless of the arc voltage. Thus, welding starts at a high voltage needed to initiate an arc, and reduces to 20 to 30 welding volts. In addition, when the stick electrode is shorted to the work at the start, the maximum current is only slightly more than the preset valve. While welding, as the arc length varies with the manipulation of the melting electrode, the current remains relatively constant. Therefore, if the arc length is varied, the voltage changes,

but the welding current, which controls weld penetration, remains close to the preset level. In addition to the power source, the stick electrode holder is the only other equipment needed to make a stick weld.

The heart of the process is the stick electrode itself. For steel welding, the center core is typically a non-alloyed steel rod. The rod is cut into short lengths, such as 14 inches. Flux ingredients are mixed with small amounts of metal alloy and a binder, often liquid sodium silicate. The dough-like mixture is extruded around the core wire. The coated rod is baked to harden the binder. A short section at the end has the coating removed, so the electrical power can be transferred to the core rod. When an arc is struck between the

Fig. 1.5. Oscar Kjellberg, founder of ESAB, invented (and received a patent for) stick electrode welding in 1904. Stick welding became more prevalent through the 1920s, 1930s, and 1940s and became the leading welding method until displaced by MIG welding. ESAB grew to be a worldwide leader in the welding field.

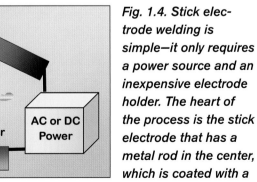

Fig. 1.4. Stick electrode welding is simple—it only requires a power source and an inexpensive electrode holder. The heart of the process is the stick electrode that has a metal rod in the center, which is coated with a mixture of flux ingredients, small amounts of metal alloy, and a liquid binder. The coated rod is baked to harden the binder.

core wire and the workpiece, the flux melts and some gaseous products, such as carbon dioxide, are formed, and this helps protect the weld puddle from oxidation and nitrogen contamination. The flux ingredients melt and form a slag that floats to the top of the weld puddle and protects it from atmospheric contamination as it cools. A variety of electrode types are available. Some can weld high-strength steels and match their strength and toughness.

Note the official AWS designation for stick welding is SMAW for shielded metal arc welding. The common term "stick welding" is used throughout this book.

MIG Welding

In 1950, Gibson, Muller, and Nelson, working at the Airco development laboratories, patented the MIG (metal inert gas) welding process. Over the ensuing decades it evolved, and today it is used to deposit more than 60 percent of the filler metal in the United States.

The MIG process utilizes a constant-voltage power supply. Similar to a car battery in output characteristics, the voltage stays relatively constant as current rises. A small-diameter solid wire, typically .030 to .045 inch, feeds from a spool through a flexible cable and MIG gun. The MIG gun has a copper nozzle that directs shielding gas to protect the weld from oxidation and nitrogen contamination. The wire exits the front of the gun through a copper contact tip where it picks up electrical power. Current flows from the copper contact tip to the small diameter wire. As current passes through the wire on the way to the workpiece, resistant heating increases the temperature to perhaps to 500 degrees F. Then an arc forms between the end of the wire and the workpiece, and as a result, the arc melts both the wire and the workpiece. Unlike TIG welding that requires careful hand manipulation to maintain the arc length, MIG arc length is maintained automatically. However, the weldor must still control the distance from the MIG nozzle to the work to achieve proper welding performance. This is covered in detail, with examples, in Chapter 6.

MIG welding can be used for steel, stainless steel, aluminum, and some other materials. The official AWS designation for MIG welding is GMAW (gas metal arc welding), but the common term MIG is used throughout this book. For those outside of the United States, the term MIG is only used when 100 percent argon or argon-helium shielding gas mixtures as used when welding aluminum. For any shielding gas that includes oxygen or an oxygen compound, such as carbon dioxide, MAG (metal active gas) is used. My purest friends cringe when I use MIG at AWS Section talks. I tell them it is far better than calling the process wire welding!

Oxyfuel Cutting

The title of this section was purposely changed from "Oxyacetylene Welding" to "Oxyfuel Cutting." The AWS term for the process is OFC (oxyfuel cutting). OAW is commonly

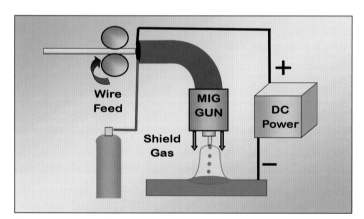

Fig. 1.6. The MIG welding process utilizes a constant-voltage power supply. The output is similar to a car battery in which the voltage stays relatively constant as current rises. A small-diameter wire, typically .035 or .045 inch, is fed from a spool through a flexible cable and MIG gun. The wire exits the gun and current flows to an arc, which forms between the wire and workpiece.

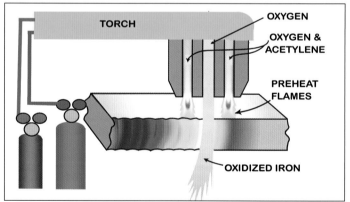

Fig. 1.7. Similar to the oxyacetylene welding torch, a cutting torch uses a special tip. Mixed oxygen and fuel gas exit multiple holes around the outer edge of the tip producing high-temperature flames. A large center hole in the cutting tip emits a high flow rate of pure oxygen, and this oxygen does the cutting. The flame preheats the steel and the oxidation of iron generates the heat to maintain the cut.

used for welding because acetylene is the only gas that can be effectively used for welding, while a number of other fuel gases can be used for cutting. In fact, natural gas can be used for automatic cutting machines.

The process for cutting steel is shown in Figure 1.7. Similar to a welding torch, a cutting torch utilizes a special tip that has multiple small holes around the outside of a larger center hole. These small holes flow mixed oxygen and fuel gas and produce the high-temperature preheat flames. A larger center hole in the cutting tip has a high flow rate of pure oxygen only, and that oxygen does the cutting.

The hot outer flames preheat the steel, and the oxygen converts the hot iron to iron oxide. This chemical oxidation generates the heat to keep the cut going! Therefore, any fuel gas can be used to start the process, but depending on which fuel gas, it may take longer to start the cutting action. However, with a steady hand, the cut can be made.

Acetylene is often the preferred fuel gas. If you don't have a steady hand, the cutting may stop with other fuel gases. When using the hotter, more intense acetylene flames

even with slight hesitation it continues cutting, and that makes it easier to operate.

I found that out by experience when I was troubleshooting a welding job and needed to take a sample of a 1-inch-thick weld on the airplane. All of the weldors and cutters were at lunch so I offered to make the sample cut. I picked up the torch and proceeded. My experience had been cutting with oxyacetylene. This torch was using propane as a fuel gas. After a number of starts and stops, the cut was made, but it wasn't pretty!

Plasma Cutting

Bob Gage, working for the Linde Development Labs (my former employer), invented plasma cutting in 1957. Welding was discussed in the patent, but the process has gained much more popularity for cutting. Initially, the process used nitrogen cutting gas; today, manual systems mostly use compressed air as the plasma gas. The process creates an arc between a non-consumable electrode in the plasma torch and the workpiece. However, unlike TIG welding, the arc is forced to go through a very small hole that concentrates the heat

and raises the temperature of the exiting plasma gas. The exiting gas in the center of the arc column reaches more than 30,000 degrees F, and that melts and blows away any conductive material. The innovative design and the rapidly swirling plasma gas in the nozzle throat allows the much-lower-melting-point copper nozzle to remain unharmed from a 30,000 degrees F arc coming through its small-diameter orifice.

When air or nitrogen are used as the plasma gas, some nitrogen compounds form in a thin area near the surface of the cut material. If the cut edges are going to be welded, they should be ground to remove this thin layer.

The AWS designation for this process is PAC, for plasma arc cutting. However, few folks use that acronym. In this book I use the shortened form, plasma cutting.

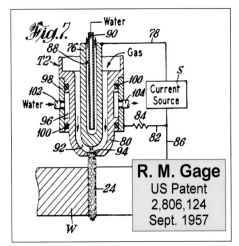

Fig. 1.9. Bob Gage, working for the Linde Development Labs (my old employer), invented plasma cutting in 1957. Welding was mentioned in the patent, but its use for cutting has gained much more popularity. Initially, plasma cutting used nitrogen as the plasma gas. Today, however, manual systems mostly use compressed air as the plasma and cutting gas.

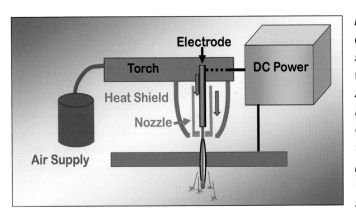

Fig. 1.8. A plasma cutting torch is similar in some ways to a TIG torch. An arc is formed between a non-consumable electrode and the workpiece. However, the arc is forced to go through a very small hole, which concentrates the heat and raises the arc temperature to more than 30,000 degrees F. That is more than five times hotter than an oxyacetylene flame, and therefore it can cut through most materials and thicknesses.

JOINT TYPES

Hundreds of joint types are used in welded fabrication. Butt joints, tubular structural joints, and fillet welds are the most common. Complex joint designs are used for welding thick sections, and for these complex joints, J- and U-grooves are used to reduce the amount of filler metal required to complete a weld. Also, many joint types are used to weld sheet metal and tubular structures, which are employed in various industries. Fabricators developed many of these weld joints as an efficient method of achieving the fit-up needed to make consistent quality welds.

Various fabrication specifications exist that define specific welding procedures and detailed welding amps, volts, and travel speed ranges for these joints. This allows a fabricator to use specific joint types without the need to prove the joint can produce the required weld quality. A number of these weld joints are shown in this chapter and may provide ideas for their use in a specific street rod or race car project.

Structural tubular joints are often used for race car chassis and roll bars. In addition, exhaust systems use thin-wall tubes that must be joined. A number of industries use tubular members for the fabrication of structures, such as building members and highway sign supports. These designs are subjected to varying loads. Designers use sophisticated stress analysis techniques to optimize the use of materials. Some of this design and welding information can be useful in race car and street rod fabrication.

Butt Joints

Simple square butt welds are often used in automotive-type welding. Variations may be useful to provide increased strength. Welding from one side only can leave some of the root area unwelded. This leaves a stress concentration that can cause a crack to form in the weld. Where maximum strength is required, a full-penetration weld should be used. TIG can produce full-penetration joint welding from one side, but you need to carefully control the penetration and be sure the full joint is melted.

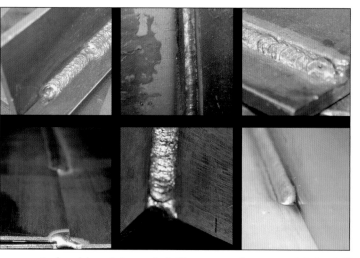

Fig. 2.1. Hundreds of joint types are used in welded fabrication. A number of more complex joint designs relate to welding thick sections, where J-grooves and U-grooves are employed to reduce the amount of weld metal needed. There are also many joints defined for use in sheet metal, such as for ductwork, that could be used in street rod applications.

Fig. 2.2. Fabricators have developed many weld joint designs as efficient methods for achieving the fit-up needed to make consistent quality welds. A number of industries that use tubular structural components have developed design criteria and weld procedures, and these can be adapted to race car and street rod fabrication.

When you can weld from both sides of the joint, a full-penetration weld is easier to accomplish. For thin material, the edges can be butted together, a weld made on one side, and a weld made on the back side that fully penetrates into the first.

A single V-preparation for butt joints is a proven method accepted for a number of design specifications and can ensure a full-penetration weld is achieved. V-preparation used for a full-penetration joint is particularly useful for thicker materials, such as 3/16- and 1/4-inch plate. By first using a single V-preparation, leaving half the plate thickness as a land accomplishes two things. It ensures good penetration on the first weld and leaves a land under the V that prevents excess penetration where the fit-up is not perfectly tight.

The V can often be made with a grinder, but you must be careful to leave half of the surface as the land.

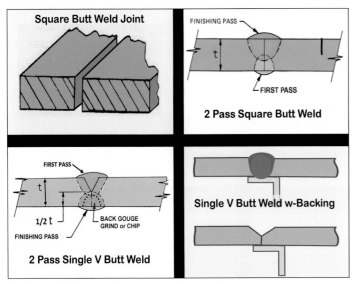

Fig. 2.3. A full-penetration weld should be used to achieve maximum strength. In the bottom left panel, the joint shown is very useful for somewhat thicker material. First, a weld is made in a single V-joint to achieve good penetration. Then the back side is gouged or ground into sound weld metal, making a U-groove. A second weld is then made in the U-groove to fully penetrate into the first deposit.

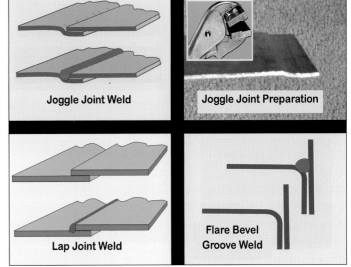

Fig. 2.4. These joints are suited for welding sheet metal. The upper left joint is commonly called a joggle joint or flange joint. The official AWS Sheet Metal Code name is an offset lap joint. Whatever it's called, this is an excellent joint when welding a patch panel. Simple locking-type pliers are available that can progressively form the edges providing a backing for the subsequent weld.

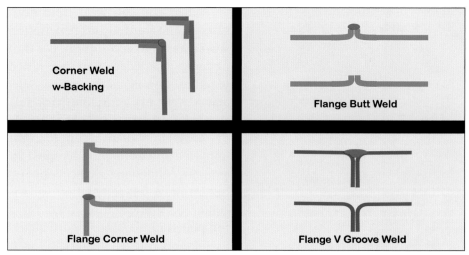

Fig. 2.5. Fabricators, including those making air handling ducts and tractor cabs, weld sheet metal. They have developed a number of joints that make it easier to weld specific sections. Some are useful for specific automotive applications. Flange joints make welding easier and may require less heat input. Backing a weld, such as the corner weld shown, adds strength and allows less precise fit-up.

The first weld is placed into the V-groove. It should be made with sufficient current and speed to penetrate about three-quarters of the plate thickness.

Then the back side is gouged or ground into sound weld metal by using a grinder held on its side or an air-powered chipper with the proper groove should go sufficiently deep, so the bottom reaches defect-free weld metal in the first-side weld, and it should result in a U-shaped joint.

A second weld is then placed in the U-groove with sufficient current, so it fuses into the groove on the first pass. The resulting weld should overlap about 20 percent of the joint thickness. This overlap eliminates any root defects that may have been created in the first weld.

A J-groove is essentially half a U-groove and can be employed where the edge of a thick plate butts to a vertical member, as might be encountered in a cross-brace attachment to a side frame rail. As with a U-groove, a J-groove minimizes the amount of

weld metal and weld heat while still ensuring adequate penetration.

Square butt welds made in sheet metal require very close fitment. Several techniques are employed to

make welding these joints easier. One approach is making a joggle or flange joint used to fabricate propane tanks, fire extinguishers, and other thin sheet metal vessel end-cap welds. With this approach, one edge of the joint is formed so the joining plate fits over the bent area. This provides back support for the weld, and it's more tolerant of slight fit-up variations.

Offset lap joint is the official AWS Sheet Metal Code name for this type of joint. The weld itself is referred to as a flare-bevel weld. Whatever you call it, this is an excellent joint when welding a sheet-metal patch panel. Simple locking pliers are available with dies welded to the grip faces, from companies such as Eastwood, that can progressively form the edges, providing a backing for the subsequent weld. There are air-powered devices that provide the same progressive crimping and make the task for preparing the panel faster.

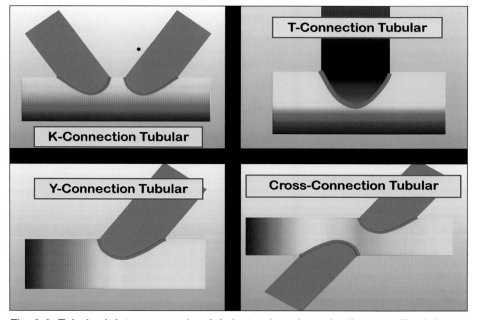

Fig. 2.6. Tubular joints are used to fabricate chassis and roll cages. The joint names shown are from the AWS Structural Welding Code for Steel. This joint type is a partial-penetration weld because there is an unwelded area at the root of the fillet weld. That creates a stress riser that can cause a crack in the fillet weld when subjected to high-stress, cyclic loading.

A number of industries fabricate sheet-metal parts such as air handling ducts and tractor cabs. A number of joints have been developed to make it easier to weld specific types of sections. Some of these designs, which include flange joints, may be useful for specific street rod applications. These flange joints, as they are referred to, make welding easier and may require less heat input. Melting the edges of a flange butt weld, as shown in Figure 2.5, is easier than making a square butt weld in sheet metal. In addition, the edges can be easily clamped together and the joint tack welded prior to final welding of the seam.

The same fit-up and welding benefits exist for the flange corner weld. Backing a weld with another part, for example in a corner weld, adds strength and is more tolerant of less-than-precise fit-up.

Tubular Structural Joints

Tubular intersection joints are typically used in race car chassis and roll cage welding, and a number of non-automotive industries use tubular members in construction. They have developed standards that define allowable loads for various intersecting tubular joints. The AWS Structural Welding Code for Steel defines the official names of these intersections. Several of these commonly used for race car fabrication are shown in Figure 2.6.

This type of joint is considered a partial-penetration weld because there is an unwelded notch at the root of the fillet, and this unwelded area creates a stress riser at the weld root. Depending on the loads involved, this stress concentration can cause a crack to propagate into the fillet welds on thinner-wall tubes,

such as those used in chassis and roll cage constriction. This is a problem with high-stress and cyclic loading. The allowable stress calculations can reduce the amount some of these joints can be safely loaded by a factor of 70 percent or more. Fatigue is a failure mode in which loads vary in a cyclic manner.

The stress riser, such as the unwelded root of a fillet, can cause a crack to form. Over time, with increasing loading cycles, these small cracks grow bigger and can lead to failure. An advantage of steel is that at a low-enough load level, the crack tip blunts and stops propagating. At that load level, the fatigue life of the structure is said to be infinite. With a fully penetrated weld or base material free from significant defects, that load or stress, to have infinite fatigue life, is about half the material's ultimate strength. However, with high-stress

concentrations, the load to achieve infinite life may be only 20 percent or less of the ultimate strength.

This infinite life characteristic is not applicable to all metals. Aluminum, for example, has no load that eliminates the growth of highly stressed cracks. For highly cyclic loading, such as a rotating member, aluminum is not a good choice.

Race cars often use many complex tube intersections for a lightweight, ridged structure. A NASCAR chassis is shown in the upper left of Figure 2.2 that has six tubes coming into one common point from various angles. To achieve the required welded-joint quality it is essential to have very good fit-up with minimum gaps. Time spent in joint preparation saves time in welding and produces the best quality structure. In Chapter 4, examples are shown of both proper and improper

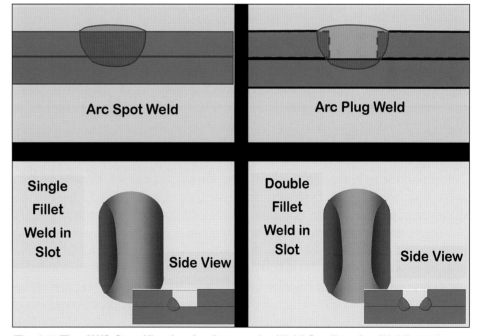

Fig. 2.7. The AWS Specification for Automotive Weld Quality—Arc Welding of Steel, defines the names for this series of sheet-metal weld joints. Arc and plug welds are commonly used for street rod fabrication. A plug weld is made through a premade drilled or punched hole. For thicker materials, a fillet weld can be made in an elongated slot.

joint fit-up. In some instances small grinding wheels or abrasive cartridge rolls may be employed to achieve the desired maximum gaps of about .010 inch for thin tube walls such as .040 inch. For .062 and thicker wall tubes, .020-inch maximum gaps should produce satisfactory welds.

Gussets can be used on tube joint intersections to stiffen the assembly. They are particularly useful for some high-strength materials such as 4130 chrome-moly, where a somewhat lower-strength, more ductile welding rod and smaller weld size can be offset with the added strength supplied by a gusset. An example of the use of a gusset is shown on a NASCAR roll cage in the lower right of Figure 2.2.

Weld Types

A fillet weld is a triangular-shaped deposit commonly used for many joints where two materials to be joined intersect at angles. In instances where two flat plates are joined there is little joint fit-up required. However, for fillet welding intersecting tubes, the complex joint geometry must be properly cut and matched to achieve the needed maximum gap of .010 to .020 inch. It is also important to ensure the bottom of the fillet weld, at the intersection of the shapes being joined, is melted and fused. It is possible to make a fillet weld having a good surface appearance that is not properly fused in this bottom area, called the root of the fillet.

Fillet welds are often partial-penetration welds and require a reduction in allowable loads because of the gap left at the weld root. The reduction factors depend on the exact joint, the amount of penetra-

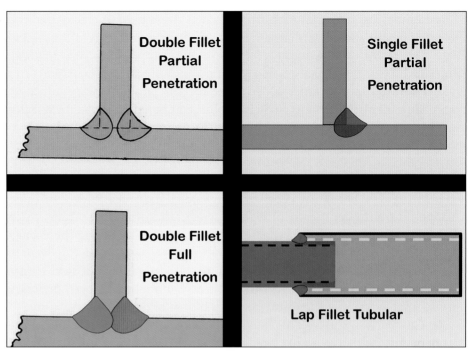

Fig. 2.8. Fillet welds are considered partial-penetration welds and require a reduction in allowable loads because of the unwelded area at the root. The amount of reduction depends on specification requirements. A single fillet has the highest stress concentration because of the loading. The double fillet is better to use, and the full-penetration double-fillet weld in which the root gap is eliminated is best.

tion, the type of loads involved, and design specification requirements.

A single fillet has the highest stress concentration because of the loading. If loaded so that the joint is bent toward the weld, the stress at the root of the weld is significantly increased.

A double fillet weld is better because, although an unwelded area exists, when a side load is applied the stresses are shared by the two fillet welds and the root stress concentration is not as high as with a single fillet.

A full-penetration double fillet is the best joint because there is no unwelded area.

The commercial automotive industry has developed its own standards for the types of welded joints. These sheet-metal weld joints are defined in the AWS Specification

for Automotive Weld Quality—Arc Welding of Steel. They include plug and spot welds, which are commonly used for street rod welding.

The difference between these two welds is that a plug weld is made through a premade drilled or punched hole while a spot weld relies on the arc melting through the top sheet and into the bottom sheet. This works well for thin sheet metal and quality can be ensured by having accurate times along with control of amps and volts.

The timing for a spot weld should start after an arc is established by having the welding machine start the timing sequence only when voltage and amperage are detected. If more strength is needed, for example on heavier top sheet materials, welds can be made in an elongated slot.

OXYACETYLENE WELDING

In Chapter 1, I mentioned oxy-acetylene welding as an ideal process to learn and practice in order to gain the fundamental knowledge of fusion welding. Melting of the base material occurs slowly and is easy to observe. Filler metal is added separately and the two-hand technique used is similar to what is employed for TIG welding. Since it is a slower process than TIG, the manual skills are easier to practice. Maintaining a fixed distance from the welding tip

to the work is not as critical as with TIG, which also helps develop manual skills.

In addition, the equipment needed is relatively inexpensive and can be used for welding, brazing, cutting, and heating. With the proper attachments, it is ideal for cutting steel of any thickness or heating metal for bending.

Therefore, although it may not be widely used in automotive work for general welding, it is good to

review the basics before discussing TIG welding (see Chapter 4). It may be the proper tool to weld a very lightweight chassis made from small-diameter 4130 chrome-moly tubing (see Chapter 7). In addition, there are some unique joining applications, such as repairing a crack in cast iron, when braze welding with oxyacetylene is often preferred. The simplicity

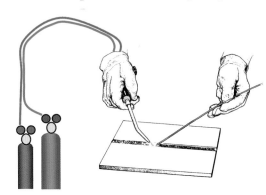

Fig. 3.2. Two cylinders, one containing oxygen and the other acetylene, supply these gases through long hoses to the torch. The gases are mixed before exiting through a small hole in the torch tip and ignited. For welding, acetylene is the only practical gas to use. (Figure adapted from ESAB's Oxyacetylene Handbook with sketch by Walter Hood)

Fig. 3.1. Oxyacetylene welding is an ideal process to learn and practice to gain the fundamental knowledge of fusion welding. The equipment is inexpensive and flexible to use. With the proper attachment, it is ideal for cutting steel of any thickness or heating metal for bending.

and relative low cost of oxyacetylene welding is another reason to consider its use for a number of applications.

Equipment

An oxyacetylene welding rig consists of two cylinders with regulators and hoses attached to a torch.

One cylinder contains oxygen and the other acetylene; they supply these gases to the torch through long hoses. Valves in the torch control the flow of gases. The gases mix before exiting through a small hole in the tip and are ignited. Different-size tips are available for welding various thicknesses of materials. For welding, acetylene is the only practical gas to use. It has the hottest inner cone temperature of any fuel gas (5,720 degrees F). Other fuel gases are acceptable for cutting because the hottest flame is not necessary.

Oxygen Regulator—Avoiding Explosions

Oxygen is usually supplied in a high-pressure cylinder with a pressure of 2,000 to 2,600 psi (pounds per square inch). This high pressure must be reduced to less than 15 psi for welding the typical material thicknesses used in automotive applications.

Connecting an oxygen regulator and adjusting the pressure must be done following the manufacturer's instructions. Be sure to read and understand the operating and safety precautions. When installing the regulator on a cylinder and opening the cylinder valve, very high oxygen pressure enters the regulator passages. This sudden high-pressure surge can cause a high temperature similar to what occurs in a Diesel engine combustion chamber when air is compressed. Most materials, including stainless steel and copper, burn and melt in the presence of pure oxygen. If not handled properly, the regulator could explode causing serious injury.

The internal design of an oxygen regulator must consider the possibilities of an oxygen fire or explosion. For these regulators to function efficiently, a valvespring is located on the high-pressure side of a diaphragm to regulate the pressure. This valvespring attaches to a valvestem and valvestem guide, which smoothes out the movement of the valvestem, so the regulator does not chatter due to rapid opening and closing of the valve.

Fig. 3.3. An oxygen regulator reduces the approximate 2,600 psi in a high-pressure cylinder to 5 to 15 psi, so it can be safely used for welding. Proper precautions must be followed when installing an oxygen regulator to avoid the possibility of a fire or explosion. Always follow the manufacturer's instructions.

Here are the steps to avoid an explosion:

1. Before installing a regulator on a cylinder, visually check the cylinder outlet to make sure there is no debris.

2. Crack open the cylinder contents valve to clear any potential contaminants from the opening.

3. A slight opening is all that is needed. Quickly close the valve.

4. The oxygen regulator must have an inlet filter, so carefully check to see that it's in place. If it is missing or not working, do not use the regulator. Take it to your gas supplier and have one installed or have the existing one repaired.

5. With an inlet filter in place, make sure it is clean, and free of all oil, grease, or contamination. This is particularly important in a location where grease or oil may be in the area where the regulator was placed when changing cylinders. Locate a clean location to place the regulator while swapping cylinders. The workbench needs to be free of grease or oil so you do not contaminate the regulator, and never put the regulator on the floor. Remember that even a small amount of oil or grease located on the regulator inlet can cause a regulator explosion.

6. After the regulator is connected, tighten the nut with no more than a 12-inch-long wrench. The sealing seats on the regulator inlet and cylinder valve are carefully machined metal-to-metal surfaces. Never use thread sealer on any regulator threads because regulator threads are not sealing threads, rather they are only for pulling the seats together. If the seats leak, the threads do not block the flow of gas. With oxygen service, not using

thread sealant is even more important because contamination of the inlet occurs.

7. If the seats are clean, they do not need high torque to properly seat. If they do need high torque, either the cylinder or regulator may have a defective seat.

8. If a leak is found after testing, remove the regulator, clean the seat with a clean cloth, and reinstall. If the leak is still present, replace the cylinder or the regulator.

The way a cylinder valve is opened is very important. A valve on any high-pressure cylinder should always be opened very slowly. One reason is the pressure gauge has a small, sealed, curved tube that bends when subjected to pressure. When the tube bends, it actuates levers that move the gauge pointer, and rapidly opening the cylinder stresses the tube.

An explosion is possible if the valve on the oxygen sensors is opened too quickly (I have witnessed many tests of oxygen regulator explosions in a laboratory environment, so this next step is reinforced in my mind). Before opening the oxygen cylinder valve, be sure the regulator pressure adjusting screw is turned out fully. Leaving the pressure adjusting screw

turned in to maintain a setting is a very dangerous practice. All regulator manufacturers warn about the need to fully turn the adjusting screw out By leaving the pressure adjusting screw in, even slightly, the high-pressure oxygen gas passes through the open valve seat, exposing the regulator diaphragm and other internal parts.

With the pressure-adjusting screw fully open, very slowly open the cylinder contents valve. One suggestion is to not look at the contents gauges while opening the valve. In fact, take an extra precaution and stand to the side of the regulator. In lab tests, when an explosion occurs, most (but not all) of the flame comes out of the front and sometimes the rear of the regulator. One reason for opening slowly is to avoid a shock to the pressure gauge, but another reason is that a regulator explosion might be caused by high pressure rapidly entering the small gauge tube that bends when pressure is applied.

If even a small amount of hydrocarbon contaminant enters the tube, and the pressure suddenly increases from 14.7 to 2,600 psi of pure oxygen, spontaneous combustion can occur. It is like a Diesel engine combustion chamber. No spark is needed, just fuel and high pressure.

A Diesel engine with a high compression ratio of 25:1 has a maximum cylinder pressure of 368 psi of air (25 x 14.7). That is sufficient to ignite Diesel fuel in air. With 2,600 psi of oxygen, it takes little fuel to cause ignition! Once the fire starts, all materials burn in an oxygen environment—the brass regulator body and even a stainless steel diaphragm.

An oxygen regulator burnout is not an everyday occurrence. There are hundreds of thousands of oxygen regulators on cylinders in the United States and just a few burns occur each year. Evidence collected over the years shows that when one does occur the inlet filter was often missing. While filters are necessary for all uses and in all environments, clean filters are especially important in auto body shops, garages, or any place where hydrocarbon products are on benches or floors.

When a full cylinder is being exchanged for an empty cylinder, the oxygen regulator might be placed on a dirty bench or floor. The inlet nipple could pick up some grease or oil and enter the opening. Burnouts have occurred in coal mines where safety precautions are always emphasized and taught to all workers. Could it be the regulators were exposed to coal dust when swapping cylinders?

Most burnouts occur when a new full cylinder is being installed. If the inlet filter is always checked before installing an oxygen regulator, the pressure adjusting screw is backed all the way out, and the cylinder valve is opened very slowly, an oxygen regulator burnout or explosion should never be experienced.

There are also oxygen regulators designed to contain an explosion should one occur. Ask your welding supplier about them.

Fig. 3.4. Acetylene regulators only allow pressures up to 15 psi because, beyond that, pressure acetylene gas is unstable. At 29 psi the gas can spontaneously explode. When acetylene is contained in a cylinder, the gas is dissolved in acetone.

Acetylene Regulator

Acetylene is potentially unstable at pressures over 15 psi. To increase a cylinder's capacity, while providing a safe environment, the acetylene is dissolved in acetone and held in a porous media contained in the cylinder. Cylinder pressures of 250 psi can then be used. The acetylene regulator lowers the pressure to the 5 to 8 psi required for welding. Consult the manufacturer's operating and safety instructions for a particular regulator and welding tip to define the pressure setting recommended. Never set pressures above 15 psi.

Torches

Most oxyacetylene welding systems are purchased as kits that include a torch handle, a cutting attachment, and several welding tips and cutting nozzles for various thicknesses of material. Some kits include a multi-flame heating head. This is similar to a large welding tip but instead of one hole with a single flame, it has multiple holes for multiple flames. This allows heating a wider area without a single hot flame melting the material.

The torch handle has two valves used to adjust the flow rate of oxygen and acetylene. These require fine adjustment to obtain the proper flame properties. However, since the oxygen valve is also used for cutting where higher flow rates are required, it must also pass a high rate of gas flow when needed. The flow rates of gas may vary from 4 to 200 cfh (cubic feet per hour), a very wide range.

The welding tip often contains a mixer where the oxygen and acetylene are combined so they can burn. Two types of mixers are in use today. The first is called a medium-pressure type, where the gases are supplied at about equal pressures. The pressures, when using medium-pressure mixers, are typically similar for oxygen and acetylene, at about 2 to 7 psi depending on the tip size. The second is an injector type, where the oxygen is supplied at a high pressure (55 psi or higher) and the acetylene is supplied at a low 1 psi.

In the injector type, the oxygen passes through a very small orifice in the injector, and the expansion of the oxygen as it leaves the orifice

Fig. 3.5. High-pressure gas cylinders, such as those used for oxygen, are made of high-strength steel and have a significant safety factor in their design. They are usually filled to about 2,500 psi, 170 times atmospheric pressure. Acetylene cylinders only contain 250 psi when full. They are filled with diatomaceous earth or a ceramic material and store the acetylene in acetone.

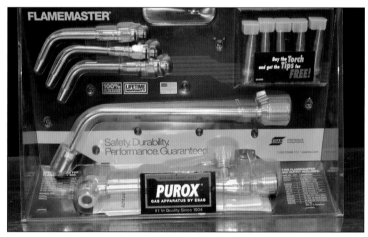

Fig. 3.6. Most oxyacetylene welding systems are purchased as kits that include a torch handle, a cutting attachment, several welding tips, and cutting nozzles for various thicknesses of material. Some include a multi-flame heating head that allows heating a wider area with flames that do not melt the material.

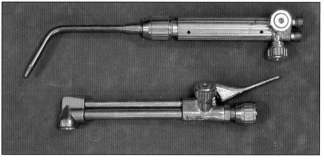

Fig. 3.7. The torch handle has two valves that are used to adjust the flow rate of oxygen and acetylene. These require fine adjustment to obtain the proper flame properties and high-flow capacity for cutting. Flow rates of gas may vary from 4 to 200 cfh.

pulls the acetylene into the mixing chamber. The mixer is located in the inlet of the welding tip because there is a relationship between the mixer and the tip size. A single mixer cannot satisfy all of the requirements.

In addition, all the passages in the welding head must be designed so that if the flame is forced back into the head, it is extinguished without damage to the head or torch. This could occur by momentary contact of the torch tip against the work. This can lead to a flashback.

One advantage of an injector mixer design is it can use the last amount of acetylene left in the cylinder. One thing to watch for when the gas level is very low is it not only pulls out the acetylene gas but the acetone as well!

Cylinders

High-pressure gas cylinders, like those used for oxygen, are made of high-strength steel and have a significant safety factor in their design. They are usually filled to about 2,500 psi, which is 170 times atmospheric pressure. Gas cylinders are periodically pressure tested and examined for damage before they are allowed to be refilled.

A common, large oxygen cylinder is about 5½ feet high and 9 inches in diameter. When full it contains about 2,640 psi and 325 cubic feet of oxygen. The physical internal volume of the cylinder is about 1.8 cubic feet, yet it holds about 325 cubic feet of oxygen. That is the volume of the gas when it exits the cylinder and is at room temperature and atmospheric pressure of 14.7 psi, which is what you pay for.

That volume of gas is proportional to the absolute pressure, which is gauge pressure plus 14.7

psi. Therefore, the gas stored in a high-pressure oxygen cylinder (2,640 gauge reading + 14.7 psi ÷ 14.7 psi) is 180 times denser than if at atmospheric pressure. Then 1.8 cubic feet of physical cylinder volume times 180 higher density = 325 cubic feet of gas at what is called STP, standard temperature and pressure.

An empty cylinder weighs about 140 pounds. It is deep-drawn from a single piece of high-strength steel or forged from billet steel. The final shape is heat treated after forming. The cylinder is pressure tested at 3,360 psi before put in service. It must also be tested at least once every 10 years while it is in service. The U.S. Department of Transportation establishes the regulations that cover construction, testing, marking, filling, and maintenance issues.

On the top shoulder of the cylinder body is the date it was put in service and the dates when it was retested. The brass oxygen cylinder valve has a threaded outlet that is machined to Compressed Gas Association (CGA) standards, and the American National Standards Institute (ANSI) accepts these standards. All oxygen regulators sold in the United States and Canada have a mating outlet fitting that conforms to these standards. The connection is designated CGA 540 and is recognized for oxygen service only.

Never use an adapter to connect a regulator to any high-pressure cylinder. Oxygen regulators are specially made for that service. Every cylinder valve is also equipped with a bursting disk, which ruptures and allows the contents to vent in the event the cylinder reaches a pressure near the test pressure, as might occur in a fire.

Acetylene cylinders are constructed differently than oxygen

cylinders. First, they are not under high pressure and only contain 250 psi when full. As mentioned, acetylene at any pressure above 15 psi is unstable and should never be used. In fact, acetylene at 29 psi becomes self-explosive, and it does not need oxygen for this explosion occur. It decomposes into carbon black and hydrogen.

How is it possible to have acetylene in a cylinder at 250 psi when the gas cannot be used above 15 psi? The gas in the cylinder is dissolved in acetone, and therefore, it does not exist as a gas within the cylinder.

The inside of the cylinder has a unique construction. It is completely filled with porous materials. Newer cylinders are filled with diatomaceous earth or a ceramic material. Older cylinders were filled with materials such as balsa wood, charcoal, and shredded asbestos. These fillers decrease the size of the open spaces in the cylinder.

Acetone, a colorless, flammable liquid, is added until about 40 percent of the porous material is filled. The filler acts as a large sponge to absorb the acetone, which absorbs the acetylene. In this process, the volume of the acetone increases as it absorbs the acetylene, while acetylene decreases in volume.

In the cylinder, there's a safety plug with a small hole through the center that is filled with a metal alloy that melts at approximately 212 degrees F or releases the contents at 500 psi. When a cylinder is overheated, the plug melts and permits the acetylene to escape before a dangerous pressure can build up. The safety plug hole is too small to permit a flame to burn back into the cylinder if the escaping acetylene should become ignited.

Hose Types

Hose for oxyacetylene welding must be the correct type. The CGA defines these grades as follows:

- Grade R and Grade RM for acetylene use only
- Grade T for all fuel gases

Why are Grades R and RM for acetylene only? For many years, the predominant fuel gas in the industry was acetylene. Acetylene has little or no adverse effects on rubber and was only used at a maximum of 15-psi pressure. Therefore, no requirements for fuel gas compatibility were initially specified. Different types of fuel gases (propane, natural gas, methyl-acetylene-propadiene, propylene, hydrogen, etc.) became popular over time, particularly for cutting. Many of these fuel gases are detrimental to certain types of rubber. With the use of these fuel gases, combining the wrong hose and fuel could lead to premature hose failure.

Grades R, RM, and T are compatible with acetylene. If a fuel gas other than acetylene is used, Grade T hose must be used.

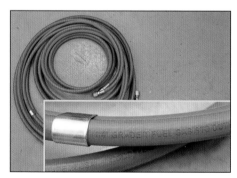

Fig. 3.8. Hoses for oxyacetylene welding must be of the correct type. The Compressed Gas Association defines three grades: Grade R, Grade RM, and Grade T. For fuel gases other than acetylene, only Grade T should be used.

Flame Types

In oxyacetylene welding, the torch tip never touches the material being welded; only the flame touches. The type of flame produced depends on the ratio of oxygen to acetylene.

A neutral flame is produced when there is a 1:1 ratio of oxygen to acetylene. This type of flame has no chemical effect on the weld metal so it does not oxidize the weld metal nor cause an increase in carbon. The excess acetylene flame is created when the proportion of acetylene in the mixture is higher than that required to produce the neutral flame. This is often called a carburizing flame.

An excess-acetylene flame causes an increase in the weld carbon content when welding steel.

An oxidizing flame is created when the proportion of acetylene in the mixture is lower than that required to produce a neutral flame. It ozonizes or "burns" some of the weld metal.

Chemistry of the Flame

When acetylene burns in the air, carbon dioxide and water vapor are the byproducts. It takes 2 cubic feet of acetylene and 5 cubic feet of oxygen or 2½ times as much oxygen as acetylene. Yet a neutral flame burns at 1:1 oxygen/acetylene ratio and a neutral flame does not have an excess of either gas. This is not a contradiction because the combustion process is more complex than simply the volume of oxygen and acetylene gas supplied to the torch. The actual combustion takes place in two stages. In the first stage, the mixture leaving the torch tip supplies the oxygen, and in the second stage, the air surrounding the flame supplies the oxygen.

In the first stage of combustion, the acetylene breaks down into carbon and hydrogen. The carbon reacts with the oxygen to form carbon monoxide, which requires one molecule of oxygen for each molecule of acetylene.

In the second stage of combustion, the carbon monoxide reacts with the oxygen from the air to form carbon dioxide. The hydrogen reacts with the oxygen from air to form water.

The two-stage combustion process produces the well-defined inner cone in an oxyacetylene flame. The first stage takes place at the boundary between the inner cone and the blue outer flame. The second stage takes place in the outer flame. If the proportion of acetylene supplied to the tip is increased, a white "feather" appears around the inner cone. This feather contains white-hot particles

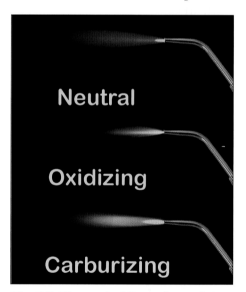

Fig. 3.9. The neutral flame has a 1:1 ratio of oxygen to acetylene. Excess acetylene causes an increase in carbon content or carburizing when welding steel. An oxidizing flame is created when the proportion of acetylene in the mixture is lower than that required to produce a neutral flame, so it ozonizes or "burns" some weld metal.

of carbon that cannot be oxidized to carbon monoxide in the inner cone boundary due to the lack of oxygen in the original mixture. On the other hand, if the proportion of oxygen fed to the tip is increased, the inner cone shortens noticeably and the noise of the flame increases.

Flame Adjustment

For most welding, a neutral flame is desired. Even a skilled oxyacetylene weldor has difficultly telling the difference between a true neutral flame and a slightly oxidizing flame. However, it is relatively easy to tell the difference between a neutral flame and a slight-excess acetylene flame. Therefore, it is always best to adjust the flame to neutral from a slight-excess acetylene flame.

Start with an excess-acetylene flame. Increase the flow of oxygen until the excess-acetylene feather is almost gone. This feather is visible in Figure 3.9, and it extends beyond the concentrated white flame cone at the torch tip.

Filler Rods

AWS classifies welding filler rods for oxyacetylene welding according to their chemical composition. AWS specification A5.2 designation of R45

is a common alloy. The rod contains low carbon and manganese alloy additions. It produces an all-weld tensile strength of about 45 ksi (1 ksi equals 1,000 psi. The use of ksi eliminates all of the zeros.)

For added strength, the R60 rod designation contains higher carbon, manganese, and some silicon.

In addition to meeting minimum chemical requirements, a weld must be made as defined in AWS A5.2 specification and produce a minimum of 60-ksi tensile strength. An even stronger alloy is available that contains other alloying elements and meets a minimum tensile strength of 65 ksi.

A note of interest is that R45 has very little alloy, which is probably similar to that of a coat hanger. For oxyacetylene welding, where only a low strength is needed a coat hanger could work. However, its chemistry may not be consistent and if you try it, be sure all of the paint or clear coating is removed.

Oxyacetylene is often more useful than other welding processes for braze welding. Unlike fusion welding, braze welding does not melt the base metal, so the melting point of the filler rod is below the melting point of the material being welded. When joining cast

iron, for example, the joint does not have to deal with a mixture of the very high carbon-based material.

A common brazing alloy with 60-percent copper 40-percent zinc has a melting point of about 1,630 degrees F, which is well below that of cast iron. Weld strength typically exceeds 45 ksi. The rod melts and wets the surface but does not melt the cast iron. A flux ensures a very clean surface and assists in the wetting process.

The other advantage of braze welding cast iron is that the high heat input and inherent slow cooling reduces the shrinkage stresses in the cast iron and avoids cracks. Braze welding is a preferred method of repairing cracks in cast-iron parts, particularly some types that are very difficult to fusion weld.

Purchasing Oxyfuel Equipment

Welding equipment can be purchased from many sources, including the Internet. However, welding equipment operates in a difficult environment and sometimes requires repair, so make sure businesses are available to repair the product if it fails.

Oxyfuel equipment is unique; it is essential to match the torches, regulators, and hose with the fuel gas. With the increasing cost of acetylene, alternative fuel gases may be best for you, and you should discuss the available options with your gas supplier. Some equipment is sold with two cylinders, but determine if your gas supplier is willing to fill them or swap them with filled cylinders instead. If you are braze welding, heating, and cutting rather than truly welding, a gas distributor may recommend a system that uses propane, propalyene, or another alternative to acetylene.

Typically Copper-Zinc Oxyfuel Braze Welding Rods				
AWS A5.8 Alloy Designation	Typical Cu copper	Typical Zn zinc	Typical Sn tin	Other Alloys
RBCuZn-C (Most common; Also available flux coated)	58%	41%	0.1%	None Significant
RBCuZn-E	58%	38%	3%	None Significant

Fig. 3.10. One area in which oxyacetylene can be very useful over other welding processes is when braze welding. The melting point of 60-percent copper and 40-percent zinc filler rod is about 1,630 degrees F, well below that of cast iron, for example. This allows a quality joint to be made without having to melt the very-high-carbon cast iron.

Safety

These general safety guidelines should be followed, but the following short overview of general safety issues is not meant to replace the instructions supplied with the oxy-fuel welding and cutting outfit or other equipment. Read and understand the manufacturer's instructions for safe use of the product.

General Precautions

- Do not use oil. Oil, grease, coal dust, and other organic ingredients are easily ignited and burn violently in the presence of oxygen. Never allow such materials to come in contact with oxygen or oxygen-fuel gas equipment. Oxygen-fuel gas equipment does not require lubrication.
- A serious accident can occur if oxygen is used as a substitute for compressed air. Oxygen must never be used to power pneumatic tools, to blow out pipelines, to dust clothing, or for pressure testing.
- Never use acetylene at pressures above 15 psi.
- Never use torches, regulators, or other equipment in need of repair. If a regulator creeps up in pressure, it has a seat leak and should be repaired or replaced.
- Do not connect an oxygen regulator to a cylinder unless it has a filter on the inlet. If it is on the input nipple, check to see it is installed every time the regulator is put on a cylinder.
- Always use the equipment manufacturer's recommended operating pressures. Using pressures higher than recommended not only makes flame adjustment difficult, but it can cause a flashback fire inside the torch.
- Always used fully enclosed goggles or a full-face helmet when working with a lighted torch. Goggles with a number-4 shade are generally satisfactory for oxyacetylene welding and oxyfuel cutting.
- Do not use matches to light a torch. Always use a friction lighter to avoid having your hands near the flame when lighting.
- Wear suitable clothing: fire resistant gauntlet gloves and long-sleeved shirts. Wool is more fire resistant than cotton or synthetic fabric.
- Before starting to weld or cut, check the area to make sure sparks, flames, hot metal, or slag will not start a fire.
- Never weld or cut without adequate ventilation.
- Use particular caution when welding and cutting in dusty or gassy locations. These atmospheres necessitate extra precautions to avoid explosions or fires from sparks, matches, or open flames of any type.
- Never weld or cut on containers that have held flammable or toxic substances until the container has been thoroughly cleaned and flammable gases have been neutralized.

Precautions for Containers with Flammable Substances
- Assume the container may contain residue.
- Wash with a strong solution of caustic soda to remove heavy oil.
- If possible, fill the container with water to within a few inches of the working area before welding. When impractical to fill with water, an inert gas such as nitrogen or carbon dioxide can be used to purge the container of oxygen and flammable vapors. Maintain the gas purge during welding.

Other Precautions
- Make sure that jacketed or hollow parts are sufficiently vented before heating, welding, or cutting. Air, gas, or liquid confined inside a hollow container expands when heated. The pressure created can cause a violent rupture.

Steel Oxyacetylene Welding Rods R45 Typically Provides 45 ksi Tensile Strength, R60 Requires 60 ksi and R65 65 ksi Tensile Strength				
AWS A5.2 Alloy Designation	Typical C carbon	Typical Mn manganese	Typical Si silicon	Other Alloys In addition to iron
R45	.06	.35	.1	None Significant
R60	.12	1.10	.20	None Significant
R65	.12	1.20	.30	.25 Cr, .20 Ni, .15 Mo

Fig. 3.11. The American Welding Society classifies oxyacetylene welding filler rods based on their chemical composition. A common alloy has an AWS specification A5.2 designation of R45. The rod contains little alloy having low carbon and manganese additions. The rod produces an all weld tensile strength of about 45 ksi.

- Remove or securely fasten in place any bushings in a casting before heating the casting. Bronze bushings expand more than cast iron when heated to the same temperature. If a bushing is left in place, the casting may be damaged or expansion may cause the bushing to fly out. If it cannot be removed, bolting large washers or plates over the ends may be possible.
- Protect cylinders, hoses, and your legs and feet when cutting. Do not cut material in such as position that allows sparks, hot metal, or a cut part to fall against a gas cylinder, the hoses, or your legs and feet.
- Take special care to make certain that a flame, sparks, hot slag, or hot metal do not reach com-

bustible material and start a fire. This is particularly important in cutting operations. Have someone stand by to watch the sparks and give warning if sparks are going into an area that could cause fire problems.
- An appropriate fire extinguisher, a pail of water, a water hose, or a sand bucket should be located and readily available near the area. A person stationed on fire watch should have firefighting equipment and a fire extinguisher immediately at hand.

Have someone remain in the area for at least a half hour after the welding or cutting is finished to watch for smoke from a smoldering fire.

Projects and Applications

This book is not intended to teach detailed manual welding techniques. Oxyacetylene welding, however is, a useful process to learn and may not be familiar to those who have some skill at MIG and TIG welding.

Project: Welding Plate Practice

EXERCISE 1: WELDING BEADS

This exercise starts with weld beads and butt welds made downhand in 1/8-inch-thick steel, and then progresses to welds made in tubing. Select a number-2 or -3 welding tip size or one recommend by the manufacturer for use with 1/8-inch-thick steel. Manufacturers also supply recommended oxygen and acetylene pressures for various tip sizes and gas mixers used with their torches. Run a weld bead on a plate. Make the first weld passes without filler metal. Hold

the inner cone about 1/8 inch from the plate and hold the torch stationary until a small molten puddle forms. Move the torch tip in a small semicircle and direct it slowly along the plate.

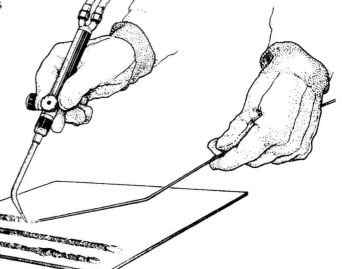

Fig. 3.12. Some simple oxyacetylene welding practical exercises help develop skills. A good exercise is to run a weld bead on plate using a 1/6-inch-diameter filler rod and hold as shown. (Figure adapted from ESAB's Oxyacetylene Handbook with sketch by Walter Hood)

EXERCISE 2: PRACTICE FORMING A WELD PUDDLE

For the next practice welds, hold a 1/16-inch rod as shown in Figure 3.12. Insert the rod into the leading edge of the weld puddle until a few drops flow into the deposit. Then pull the rod back slightly and advance the torch a small amount, allow a puddle to form, and repeat the process of inserting the rod.

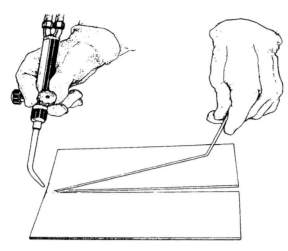

Fig. 3.13. To fill gaps you need two pieces of steel 9 inches long and 1/8 inch thick. Place them 1/16 inch apart at one end and 3/16 inch apart at the other end. Make a tack weld at the narrower end first and then tack weld the larger gap. Advance the torch tip as the weld progresses. (Figure adapted from ESAB's Oxyacetylene Handbook *with sketch by Walter Hood)*

EXERCISE 3: ADD FILLER WELD BETWEEN TWO PLATES

Filling gaps is the next thing to learn. Place two pieces of steel approximately 9 inches long, 3/32 to 1/8 inch thick, 1/16 inch apart at one end and 3/16 inch at the other. Make a tack weld at the tighter end and tack weld the end with the 3/16-inch gap. The next tack weld is somewhat more difficult because of the greater gap. Move the flame slowly from one edge to the other. As the edges melt, place a small amount of filler metal on the edge and allow it to cool slightly. This is usually best performed when the flame is on the opposite side of where the metal is added. Increase the size of the tack by adding more filler rod.

After tack welding the piece, start at the tighter end and add filler as you did for the tack; move the flame in a small arc from side to side. Advance the torch tip as the weld progresses.

Figure 3.14 is a view of the middle of the joint. The technique becomes obvious as the weld progresses. More time is spent at the edges as the torch is moved side to side. Too much time spent in the middle of the joint causes the metal to burn through. The end objective is to have the weld bead drop through slightly at the bottom while being reinforced about 1/16 inch on top.

Test the weld quality by cutting a section about 3 inches long and placing one plate in a vice and striking the other with a hammer. You should be able to bend it at a 90-degree angle without breaking it.

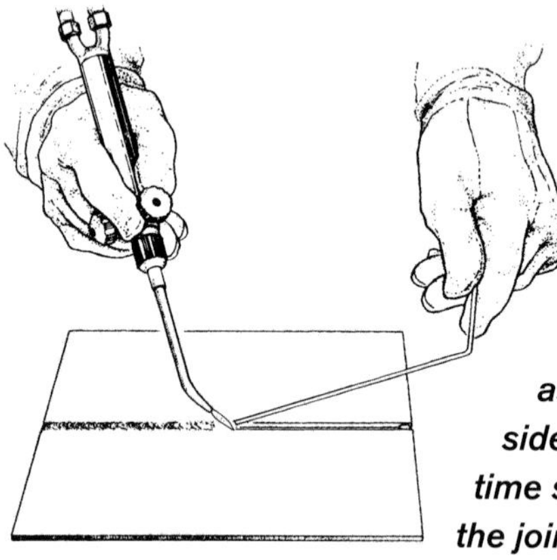

Fig. 3.14. This is the perspective in the middle of the joint. The technique becomes obvious as the weld progresses. More time is spent at the edges as the torch is moved side to side. Too much time spent in the middle of the joint causes the metal to burn through. (Figure adapted from ESAB's Oxyacetylene Handbook *with sketch by Walter Hood)*

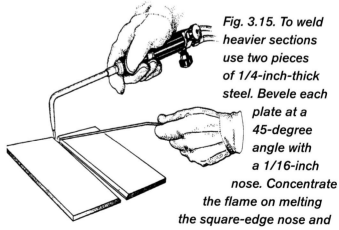

Fig. 3.15. To weld heavier sections use two pieces of 1/4-inch-thick steel. Bevele each plate at a 45-degree angle with a 1/16-inch nose. Concentrate the flame on melting the square-edge nose and forming the underbead. (Figure adapted from ESAB's Oxyacetylene Handbook *with sketch by Walter Hood)*

EXERCISE 4: ESTABLISH ROOT WELD

Heavier sections can be welded using a similar technique (but with the appropriate tip size and filler rod). Use two pieces of steel 1/4-inch thick. Bevel each plate at a 45-degree angle. Use a grinder to make a flat nose at the bottom about 1/16 inch in width. This leaves a 90-degree bevel 3/16 inch deep. Assuming the test plates are about 9 inches long, gap them 1/8 inch at one end and 3/16 inch on the other.

Make tack welds on both edges as was done for the thinner sheet metal. Once the tack welds have been made, make a weld as before, but it should only fill up about half the thickness. This is the first of two weld passes. Focus on making a good underbead. The top edges of the plate should not have melted. The flame is concentrated on melting the square-edge nose and forming the underbead.

As the root pass approaches completion, the top corners of both pieces should still be sharp and unmelted. This first pass achieves complete root penetration and little else. The appearance of the top of the weld is not very important because it will be melted and covered by the finish pass. The puddle can be kept relatively small, which is also a help if the gap varies.

Make sure the bottom-squared nose is brought up to melting temperature before the weld puddle advances across them. Be sure not to add too much filler metal at one time.

Move the filler rod in and out of the weld puddle to control the rate of filler addition.

Move the flame across the joint, spending more time on the sides of the V, being sure they are melted up to about halfway up to the top edge of the plate.

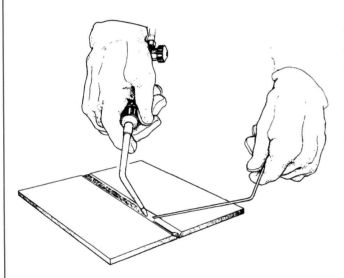

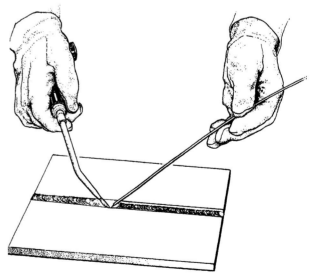

Fig. 3.16. The objective of the first pass is to complete root penetration (here, the root pass is approaching completion). Keep in mind the appearance of the top weld is not very important. Move the flame across the joint, spending more time on the sides of the V so they are melted up to about halfway. (Figure adapted from ESAB's Oxyacetylene Handbook with sketch by Walter Hood)

Fig. 3.17. With the second pass, keep the end of the rod in the puddle at all times. Rub the end against the solid metal below the puddle, maintaining a constant back-and-forth motion across the joint. When the flame is melting one side of the joint, the rod is pushing on the puddle on the opposite side. (Figure adapted from ESAB's Oxyacetylene Handbook with sketch by Walter Hood)

EXERCISE 5: COMPLETE FINISH WELD

Make a second pass over the first one. A larger 1/8-inch-diameter rod can be used as the weld progresses slowly and more filler rod is needed. Start at the beginning and form a weld puddle and proceed to fill the joint completely.

Much more filler is required in this pass. Keep the end of the rod in the puddle at all times, actually rubbing the end against the solid metal below the puddle and keeping it in a constant back-and-forth motion across the joint. The flame should move in longer arcs than was needed for the root pass. It should dwell at the sides of the V so they are melted before the puddle moves forward. Use a sideways motion of the filler rod to move the puddle back.

When making this pass, heat from the weld puddle, not the flame, melts the filler rod. The flame and rod motion must be controlled because the flame is melting one side of the joint and the rod is pushing on the puddle on the opposite side. The width of the flame movement need not be as large as the rod movement. If the rod is not moved a sufficient amount, a weld undercut occurs. There may also be places where the weld bead did not reach the top surface of the plate, creating an underfill defect.

In Figure 3.18, a backhand technique (also called drag) is used for the second pass. The flame is angled toward the completed weld, and the rod is angled toward the finishing end. Although most oxyacetylene welding is done with a forehand or push technique (as shown in all previous illustrations) in some situations, a backhand technique may be desired. The rod and flame move in an oval pattern along the weld line. The rod moves backward as the flame moves forward. The finished weld has heavier ripples than a good forehand weld deposit.

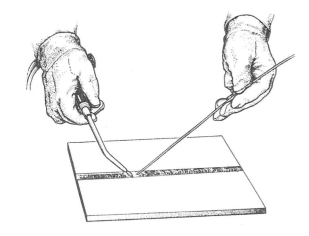

Fig. 3.18. Make the second pass with a drag, or backhand, technique. Angle the flame toward the completed weld and the rod toward the finishing end. Use a forehand technique for most oxyacetylene welding, but in some situations, a backhand technique is useful. (Figure adapted from ESAB's Oxyacetylene Handbook *with sketch by Walter Hood)*

Project: Welding Pipe Practice

Although this series of welds shows welding larger-diameter pipe, it can be used for welding small-diameter tubes. The oxyacetylene welding process is still used to weld 4130 tubing for aircraft. My old friend Butch Sosnin taught sprinkler pipe weldors that oxyacetylene welding was ideal for small-diameter pipe welding, so it certainly should be considered.

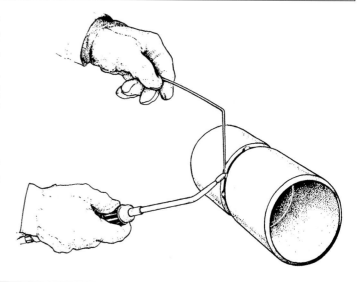

Fig. 3.19. Welding a pipe is another useful exercise. The oxyacetylene welding process is still used to weld 4130 tubing for aircraft. An application of welding 1/2-inch-diameter 4130 tubing is discussed in Chapter 7. (Figure adapted from ESAB's Oxyacetylene Handbook *with sketch by Walter Hood)*

EXERCISE 1: MAKE BOTTOM-TO-TOP WELD

Place tacks in the joint, as shown in Figure 3.20. With material thickness less than 1/8 inch, you can make the weld in one pass. Point the torch flame approximately toward the centerline on the pipe. Hold the filler rod tangentially to the joint. Make the weld from bottom to top, adding filler to the leading edge of the puddle. Completing the weld requires welding in the flat, vertical, and overhead positions. Stop the weld, repositioning your hand and start a puddle for each position.

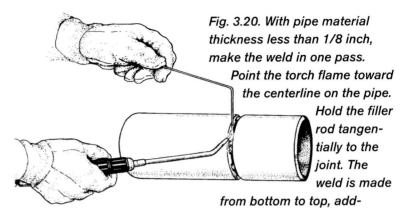

Fig. 3.20. With pipe material thickness less than 1/8 inch, make the weld in one pass. Point the torch flame toward the centerline on the pipe. Hold the filler rod tangentially to the joint. The weld is made from bottom to top, adding filler to the leading edge of the puddle. (Figure adapted from ESAB's Oxyacetylene Handbook *with sketch by Walter Hood)*

EXERCISE 2: MAKE TOP-TO-BOTTOM WELD

In Figure 3.21, the weldor is using a backhand technique, and the weld is made from top to bottom. There are certain situations in which welding out of position is necessary. In some welding circumstances, the flame is used to hold up the weld puddle. The aim is to keep the puddle from running onto metal that has not melted. As noted when backhand welding flat plate, the relative movements of the rod and flame are different from forehand welding. The rod and flame move in an oval path with one forward when the other is backward and vice versa.

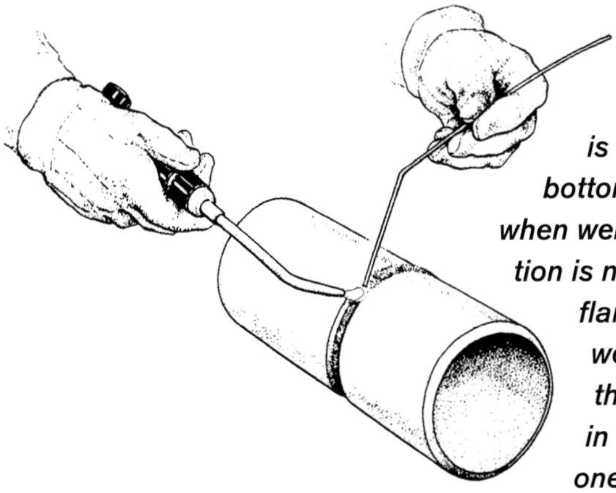

Fig. 3.21. Use a backhand technique as the weld is made from top to bottom. There are times when welding out of position is necessary. Use the flame to hold up the weld puddle. Move the rod and flame in an oval path with one going forward and the other going backward. (Figure adapted from ESAB's Oxyacetylene Handbook *with sketch by Walter Hood)*

EXERCISE 3: PERFORM OVERHEAD WELD

Welding overhead is not much different from welding vertically. Surface tension holds the molten metal up, as does the force of the flame. If forehand and backhand techniques are an option, a backhand technique can be used to start the weld at the top and move downhill. When the bottom is reached, a forehand technique is used to weld to the top. If you weld with only a forehand technique, start at the bottom and work up one side, change position, and work up the other side.

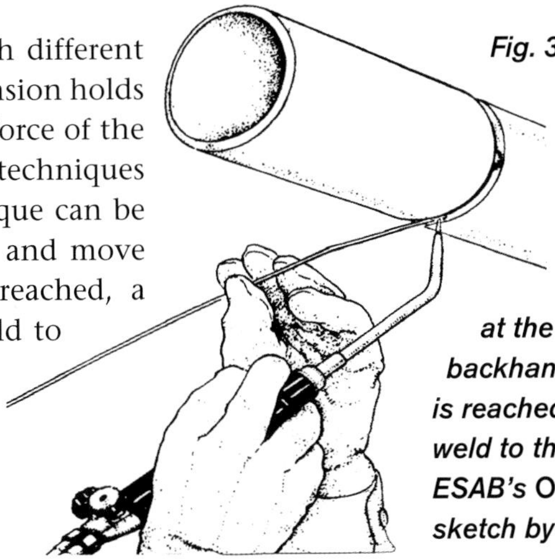

Fig. 3.22. Welding overhead is similar to welding vertically. Surface tension and the force of the flame hold up the molten metal. If forehand and backhand techniques are used, the weld can be started at the top and welded downhill with a backhand technique. When the bottom is reached, use a forehand technique to weld to the top. (Figure adapted from ESAB's Oxyacetylene Handbook *with sketch by Walter Hood)*

Application: Oxyacetylene Welding Cast Iron

Oxyacetylene welding is very effective for repairing cracks in cast-iron parts. Some types of cast iron can be fusion welded, although with difficulty and the potential for cracking, while others cannot. However, they all can be braze welded. With this approach, the filler material melts and wets the cast iron, but that occurs well below the melting point of the cast iron.

Fig. 3.23. Oxyacetylene welding is very effective for repairing cracks in cast-iron parts. Some types of cast iron can be fusion welded, although with difficulty and the potential for cracking, while others cannot. However, they can all be braze welded.

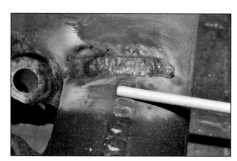

Fig. 3.24. The finished brazed joint is sound and well wet into the cast iron. The brazing material is also very ductile. When the cast iron expands and cools as the part is heated and cooled, the brazing alloy can expand and contact as needed and does not excessively stress the cast iron.

Repairing an exhaust manifold is a typical example of this.

To simulate a crack repair job, a groove representing a crack was placed in the manifold. The weldor ground through the full thickness of the manifold and a V-joint was placed in it. In addition to the advantage of braze welding not melting the cast iron, it also requires heating the material to a high temperature. This essentially preheats the assembly and avoids high cooling rates. This high heating reduces the stress on the braze weld area when completed, so the brittle cast iron is less likely to crack.

AWS RBCuZn-C (58-percent copper and 41-percent zinc), a common brazing alloy, is used to braze weld

Fig. 3.25. Braze welding does not melt the cast iron, and also requires heating the material to a high temperature. This essentially preheats the assembly and avoids high cooling rates. Stress on the joined area is reduced as a result, so the brittle cast iron is less likely to crack.

this simulated crack. The white outside coating on the rod is a brazing flux, which provides good wetting of the cast iron. To further promote wetting of the cast iron, brazing alloy can butter the surfaces to be joined and this bridges the joint. In addition, it allows the wetting to be seen more clearly. The flux is applied to remove oxides from the cast-iron surface and promote wetting. There is sufficient coating on the rod so there's no need for adding more. The brazing flux is available in a metal can. It's applied to the joint or by dipping the bare rod into the flux as the braze welding progresses.

The finished brazed joint is sound and wet into the cast iron. The brazing material is also very ductile. When the cast iron expands and cools as the part is heated and cooled, the brazing alloy can expand and contract as needed and does not excessively stress the cast iron. To be safe, the part is slow cooled after brazing.

Fig. 3.26. AWS RBCuZn-C, a common flux-coated brazing alloy, was used to braze weld this simulated crack. A flux is needed to ensure good wetting of the cast iron. It removes oxides from the cast-iron surface and promotes wetting. There is sufficient coating on the rod to avoid the need for adding more.

TIG WELDING

Russell Meredith applied for a patent in January 1941 for what became the invention of TIG welding. He worked for Northrup making lightweight military airplanes. The company needed an improved welding process for joining lightweight magnesium and aluminum.

The focus of the patent was welding magnesium because it is 35-percent lighter than aluminum and 75-percent lighter than steel. Helium was found to provide a good protective gas shield and the patent examples mentioned the use of that gas. Northrup named the resulting process and associated products Heliarc. However, the patent claims covered welding any metal with the use of an inert gas shield. The Linde Division of the Union Carbide Corporation (UCC) that manufactured and distributed industrial gases and welding equipment purchased this basic process patent. Unlike Northrup, which built airplanes, Linde had the incentive to develop an extensive range of TIG torches and other accessories to promote the use of the process for a wide variety of applications.

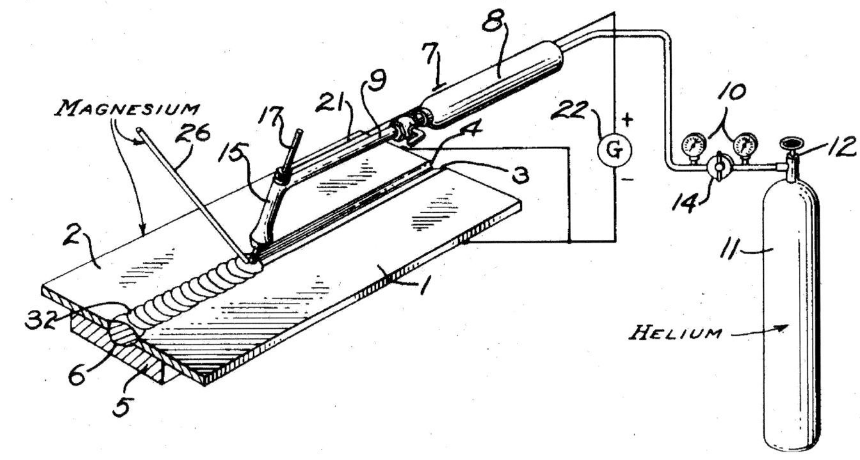

Power Options

TIG welding was initially performed with stick welding power, which was readily available at the time of its introduction. However, for welding thin materials with lower current, more precise power supplies were needed. Special power supplies were developed that were similar to stick welding power so volt amperage curves maintained a relatively constant current over the range of operating voltage. This allowed variations in arc length while still controlling the key welding parameter, current.

Fig. 4.1. Russell Meredith invented TIG welding while working for Northrup, a manufacturer of military airplanes, in 1941. It needed an improved welding process for joining lightweight magnesium and aluminum. Magnesium is 35-percent lighter than aluminum and 75-percent lighter than steel. The focus of his patent, shown here, was welding these materials with helium. The resulting process and associated products are called Heliarc.

DCSP Power

A DC straight-polarity (DCSP) power configuration, also called DC electrode negative (DCEN), is used for most steel TIG welding. In DCSP TIG welding, electrons flow from the tungsten electrode through the arc to the workpiece. Balancing their negative charge, positively charged gas ions flow from the workpiece to the tungsten electrode. A large amount of energy is required to get the electrons into the work surface. This surface energy release causes heating of the workpiece and provides high weld penetration.

DCRP Power

Another option is to use DC reverse-polarity (DCRP) power configuration or DC electrode positive (DCEP). The electrons flow from the workpiece, through the arc, to the tungsten electrode. Balancing their negative charge, positively charged gas ions flow from the tungsten electrode to the workpiece. DCRP requires a large amount of energy to drive the electrons into the tungsten electrode tip, and as a result, a large amount of heat is released into the tungsten. To prevent excess melting of the tungsten, the maximum allowable current is much lower if DCRP is used. As an example, where a 3/32-inch-diameter tungsten electrode can weld with up to 250 amps with DCSP, when DCRP is used the maximum current is 30 amps. Also with DCRP, energy released at the workpiece creates a shallow, wide melting area.

One unique feature of DCRP is especially useful for welding aluminum. When the electrons leave the work surface, they lift the surface oxides that form very quickly on aluminum. This is thought to be aided by the impingement of the gas ions on the work surface. However if DCRP is used, the tungsten electrode size has to be increased significantly to handle the maximum allowable current, even for low-current welding of sheet metal. DCRP is seldom utilized in practice.

AC Power

AC power is essentially a combination of DCSP and DCRP. With conventional power systems, the arc current switches direction between the two polarities twice during each power cycle, or 120 times per second. The arc extinguishes and reignites with current flowing in the opposite direction. This is a good compromise for aluminum where the cleaning cycle exists for half the time. A full-half cycle of DCRP is more than sufficient for the amount of oxide cleaning needed to make quality welds. However, the oxide layer makes it difficult for the arc to reignite as it passes through zero. This significantly reduces the amount of

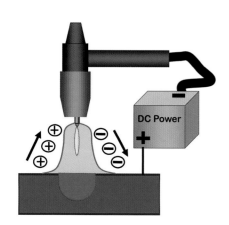

Fig. 4.2. Steel TIG welding is typically performed with DCSP power configuration. In DCSP TIG welding, elections flow from the tungsten electrode through the arc to the workpiece. Balancing their negative charge, positively charged gas ions flow from the workpiece through the arc to the tungsten electrode.

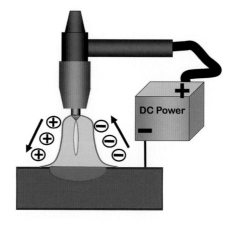

Fig. 4.3. DCRP could also be used in steel welding. The elections flow from the workpiece through the arc to the tungsten electrode. With DCRP, a large amount of energy is required to get the electrons into the tungsten. This causes a great deal of heat to be released at the tungsten tip. Therefore, this configuration is seldom used.

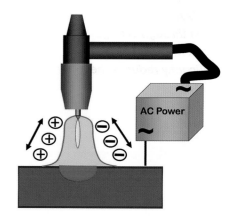

Fig. 4.4. AC power is a combination of DCSP and DCRP. With conventional power systems, the arc current switches direction between the two polarities twice each power cycle or 120 times per second. The arc extinguishes and reignites with current flowing in the opposite direction. For aluminum, during the DCRP cycle, the surface oxide that rapidly forms is removed, providing quality welds.

time spent in the DCRP mode. When these arc outages occur, the amount of cleaning is reduced and the DCRP cycle can even be completely blocked.

To provide a means of quickly reigniting the arc, a high-voltage, low-current source, similar to a capacitive discharge car ignition, can be used. On older-type conventional sine-wave AC power supplies this high-voltage, high-frequency source is usually operated continually. This ensures the arc remains on for the full cycle. This high-voltage device is also useful for starting the arc without having to touch the tungsten to the workpiece.

Newer TIG power supplies utilize components that create an AC square-wave shape rather than a classic sine-wave shape shown in Figure 4.7. Depending on the design, this may allow AC welding without the use of a continuous high-voltage, high-frequency circuit.

Modern inverter-based power supplies create their own frequency to operate and are not tied to the input power line frequency. These inverter based units operate at more than 20,000 cycles per second rather than 60-cycles on a conventional power line frequency. They can easily shape the power to be a square wave rather than a simple sine wave. This provides much better control of the arc.

Fast switching of DC from DCSP to DCRP simulates AC. Unlike

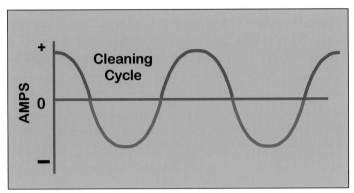

Fig. 4.5. This is a graph of AC current versus time. The oxide layer makes it difficult for the arc to reignite as it passes through zero going to the reverse-polarity half cycle. It may take the arc a short time to reignite, which significantly reduces the amount of time spent in the DCRP mode.

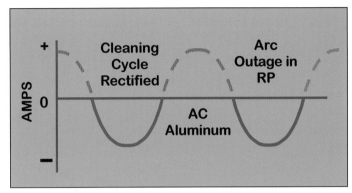

Fig. 4.6. The surface oxide layer can completely block the DCRP half cycle. When these arc outages occur, the amount of cleaning is reduced and the arc becomes less stable. Therefore, the advantage of using both DCSP for increased weld penetration and the DCRP for cleaning is eliminated.

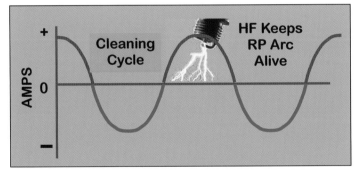

Fig. 4.7. When the welding current crosses the zero line, superimposing a high-voltage, low-current source, similar to a capacitive discharge car ignition, helps to quickly reignite the arc. For older conventional sine-wave AC power supplies, this high-voltage, high-frequency source is usually operated continually. The high-voltage approach is also useful for starting the arc without having to touch the tungsten to the workpiece.

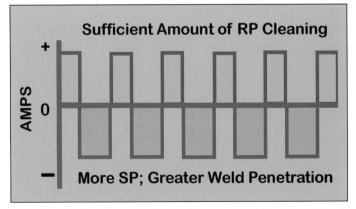

Fig. 4.8. Modern inverter-based power supplies operate at a frequency over 20,000 cycles per second. They can easily shape the power to be a square wave, providing much better control of the arc and eliminating the need for high-frequency arc reignition. The amount of time spent in the DCSP and DCRP cycles can also be independently varied.

classic sine-wave power, where the arc must reignite twice each cycle, most inverter-based square-wave power sources switch so quickly they eliminate the need for high frequency to prevent rectification. Figure 4.8 shows a typical square-wave current graph where the DCSP and DCRP are also on for different lengths of time. In this instance, more DCSP provides greater weld penetration. (See "Equipment" on page 36.)

Torch Position

The position of the TIG torch relative to the workpiece is important in controlling the welding start and duration. It also controls shielding quality. In some weld configurations, the TIG torch cup cannot reach into a crevice, such as with some intersecting tubular joints. The shielding gas flow must be limited to avoid turbulence if a small gas cup is used to increase joint access.

The following is a discussion of the methods of compensating for these issues.

Starting the Weld

Most power supplies designed for TIG welding have a high-voltage device that acts like a car ignition and sends high-voltage current between the tungsten and the workpiece. When a high-voltage start system is used, the tungsten is held about 1/8 inch from the workpiece. With the torch approximately perpendicular, the power is tuned on. The electrons jump the gap, and the arc is established. Most TIG power systems also have a control circuit, so that a foot pedal or thumb control switch located on the torch handle is used to adjust the welding current level.

A foot control is commonly used and operates similarly to the gas pedal in an automatic-transmission car. When the pedal is pushed, shielding gas preflow starts and the welding process begins. The gas solenoid is opened for a fixed amount of time (preset on the welder) and the welding power is turned on. The power supply is energized in conjunction with the high-voltage start system. The amount the foot pedal is depressed adjusts the starting current, and therefore, as the pedal is depressed, the current increases just as a car gas pedal increases speed.

Some modern systems start on low voltage and do not require high-voltage devices. They operate by touching the tungsten to the workpiece, employing a low-level signal current. When the tungsten is raised off the workpiece, the main welding power is initiated.

Other welders have controls, called four-stroke or something similar, that operate like a motorcycle transmission in which one operation sequences through the gears. A button on the TIG torch handle is used to cycle through an operating sequence in order. When the button is first depressed, the curent starts the preflow, and after a preset time, the power energizes. Pushed a second time, the current increases to a preset level. With another push, the current starts to decrease. With a final push of the button, the welding ends, and the postflow starts.

Once the arc is started, the torch is held about 1/8 inch from the workpiece and can be rotated slightly in

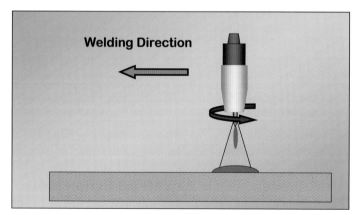

Fig. 4.9. Once the arc has been started, hold the torch about 1/8 inch from the workpiece and rotate slightly in a small circle to form a weld puddle. Tilt the torch about 75 degrees from the workpiece or about 15 degrees from a vertical position with the electrode pointing in the direction of travel.

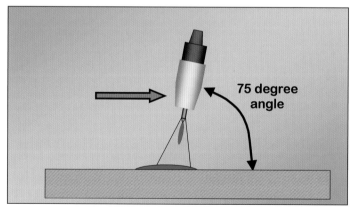

Fig. 4.10. Move the arc forward in the direction of desired travel while watching the leading edge of the puddle to see that the desired melting of the workpiece is occurring. Maintain the torch height at a fixed distance from the workpiece while along the weld joint. If filler metal is added, move the torch slightly back to the rear.

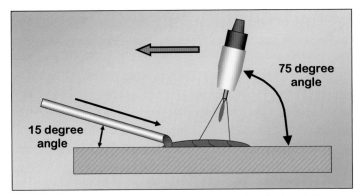

Fig. 4.11. Hold the filler rod at a low angle of about 15 degrees from the horizontal. Move the arc slightly back along the direction of travel as the filler rod is inserted into the leading edge of the molten puddle. After a drop of metal enters the weld, move the rod quickly back.

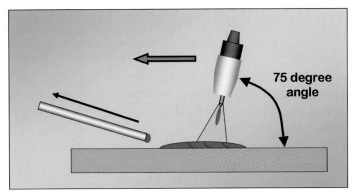

Fig. 4.12. Move the torch to the leading edge of the weld puddle. When the puddle is bright and shiny, repeat the process of moving the torch toward the rear of the weld puddle and adding another drop of metal. The added metal cools the weld puddle.

a small circle to form a weld puddle. The torch is tilted about 15 degrees from vertical so the electrode is pointing in the direction of travel.

Moving Forward

The arc is moved forward in the direction of desired travel while the leading edge of the puddle is watched, so that the desired melting of the workpiece is occurring. The torch height must be maintained at a fixed distance from the workpiece as it is moved along the weld joint. Several methods can be employed to maintain the proper torch height and proceed smoothly along the weld joint:

- The elbow of the hand holding the torch can be braced on the

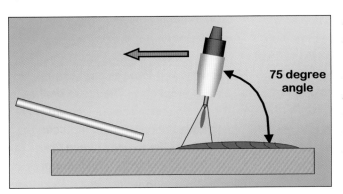

Fig. 4.13. Pull the rod away and move the arc to the leading edge of the weld puddle. The arc heat is focused on the leading edge of the puddle and observed to be sure the base metal is melting.

workpiece and the arm slid parallel to the weld joint.

- A block of wood can be used to support the arm or wrist as it is slid along the weld joint.
- The finger-drag technique uses an 8-inch-long fiberglass sleeve slipped over a glove finger in the torch hand. It is folded back and the doubled fiberglass sleeve insulates the gloved finger from the hot workpiece.
- A technique called "walking the cup" positions the tungsten extension from the cup, so that when the cup touches the workpiece, the proper arc gap is maintained. For a root weld, in a pipe V-joint, the cup can be rocked from side to side so it alternately touches both sides of the weld joint.

Whichever torch manipulation technique you choose, it should be tested and practiced with the power off to ensure the degree of control needed to make the weld is achieved. The manual skill required to maintain the close gap from the tungsten electrode to the weld puddle requires practice to achieve quality weld results.

Adding Filler Metal

The filler rod is held at a low angle of about 15 degrees from horizontal. The arc is moved slightly back along the direction of travel as the filler rod is inserted into the leading edge of the molten puddle. After a drop of metal enters the weld, the rod is quickly moved back.

The torch is moved again to the leading edge of the weld puddle. When the puddle is bright and shiny, the process is repeated, moving the torch toward the rear of the weld puddle and adding another drop of metal.

Equipment

TIG equipment has changed significantly in the past few years. Inverted-based power supplies have made significant advancements in providing improved weld performance with modern electronics. It is easier to make quality and visibly appealing welds with these modern power sources. The following discusses some of the these advances.

Conventional Transformer Power Supplies

Conventional stick welding power can be used for TIG welding. In some situations, DC stick power is used for welding root passes in carbon-steel pipe, and it's followed by stick welding for the fill and cap

Fig. 4.14. Conventional stick welding power can be used for TIG welding. However, simple AC stick-welding transformers do not work for welding aluminum because rectification of the DCRP cycle needed for cleaning the aluminum surface reduces or eliminates the DCRP half cycle. Older TIG power can make good-quality welds but does not have the advanced features of the more modern inverter-based TIG welder.

weld passes. For fieldwork, some weldors use a long cable assembly on a TIG torch with an integral manual gas valve. The torch is connected to a DC-engine-driven welder.

A special TIG power source that has starting circuits and variable-current control is needed for welding automotive-type thin-tubing or sheet-metal joints. Also, AC power is essential for welding aluminum. Simple AC stick welding transformers do not work because rectification of the DCRP cycle needed for cleaning the aluminum surface reduces or eliminates the DCRP half cycle as shown in Figure 4.6.

TIG add-on boxes are available that include a high-frequency gen-

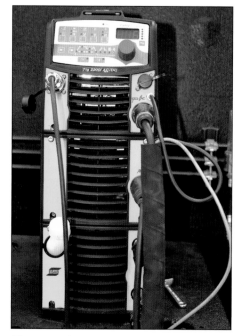

Fig. 4.15. Inverter power supplies convert the incoming AC power line power to DC. The DC power is then converted back to AC at greater than 20,000 cycles per second, allowing use of much smaller transformers. Although a conventional AC/DC TIG welder weighs more than 200 pounds, a similar-capacity inverter-based AC/DC full-featured TIG welder weighs only about 35 pounds.

erator, and some have gas solenoids. However, the simple buzz-box-type AC stick welder (the red one in Figure 4.14), does not have easily adjustable current control or the ability to use a foot-pedal control. A MIG welder may be the better choice than adding a TIG box for occasionally welding aluminum. It's not worth the money to convert this type of power supply to a TIG welder. TIG requires a separate, pure argon, shielding gas cylinder.

A number of AC/DC electronically controlled TIG welders are available. These can make excellent welds but do not have all the advanced features of the more modern inverter-based TIG welders. When shopping for a TIG welder, make sure it has a foot-pedal control and pulsing capability. Pulsing the current from a low background to a higher value makes welding thin material much easier. Utilizing the pulse control makes it easier to achieve a stack-of-dimes weld appearance and is a big help in avoiding burn-through on thin material.

Fig. 4.16. This microprocessor-based AC/DC TIG welder provides a very user-friendly interface to set the many welding parameter options. The two arrow buttons (on the left) allow cycling through and viewing the set-up procedure. The LEDs light up for the specific variable whose value is shown digitally and can be set by turning the knob under the display (on the upper right). Shielding gas preflow is shown here at 7.5 seconds.

Modern Inverter TIG Welders

Inverter power supplies operate by converting the incoming AC power line, which operates at 60 cycles per second, to DC. The DC power is then converted back to AC, but at a frequency typically greater than 20,000 cycles per second. The higher frequency has a number of advantages, one of which is the physical size of the large transformers in conventional welders can be much smaller. Transformer size is inversely proportional to the AC frequency. Therefore, 20,000 (versus 60) cycles per second allows very small transformers, with much less copper and iron. While a conventional AC/DC TIG welder weighs more than 200 pounds, a similar-capacity inverter-based AC/DC, full-featured, TIG welder weighs only about 35 pounds.

An inverter-based TIG welder switches rapidly from DCSP to DCRP power to produce AC output. The DC

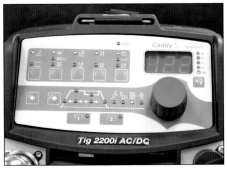

Fig. 4.19. A significant advantage of inverter-based power is the ability to increase frequency from the normal-line 60 cycles per second. Higher frequencies produce a narrower, more penetrating arc. On thinner material, such as sheet metal, about 125 cycles per second often delivers superior results by allowing lower current and therefore lower heat input when making a particular weld.

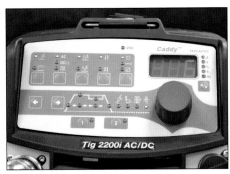

Fig. 4.18. A unique feature of the microprocessor-controlled inverter TIG welder is tungsten preheat. Rather than the electrode instantly going from room temperature to 6,000 degrees F, if preheated, starting is easier and tungsten life extended. This control sets the percentage of the weld current (in this case 6 percent) that is used in the preheat cycle. Touching the tungsten to the workpiece starts the preheat function.

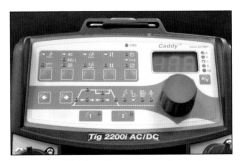

Fig. 4.17. With AC power selected (second light from the left on the top row), the next variable to set is balance. This allows setting the ratio of DCSP and DCRP cycles. The number shown, 76, is the percent of DCSP, which leaves 24-percent DCRP for cleaning aluminum. That is typically sufficient and allows the higher penetration achievable with the larger amount of DCSP.

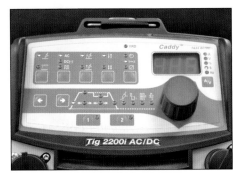

Fig. 4.20. The next variable to set is the upslope time. This is set in seconds and defines the time it takes to increase from a low starting current to the preset welding current. In this example, the time is set at 1.5 seconds.

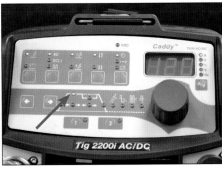

Fig. 4.21. The welding current can be set as high as the 220-amp limit. Note the LED on the line part of the display (arrow) that indicates that the high-current level is being adjusted. In this example, turning the rotary knob under the digital display sets 120 amps. This current level can be set from a low value of 3 amps.

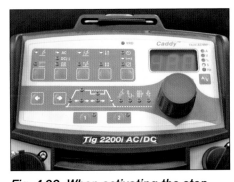

Fig. 4.22. When activating the stop sequence using a torch switch or foot control, the current reduces to 3 amps during the time set for downslope. In this instance, 2 seconds is set, which allows the torch to be held stationary or back-stepped slightly to reduce the depression in the weld puddle caused by arc force. Just prior to beginning the stopping sequence, add the last drop of filler then pull the rod away.

power reverses from DCSP to DCRP almost instantaneously so the arc does not extinguish. Therefore, there is no need for continuous high frequency to keep the arc ignited. AC current is a square wave, not a sinusoidal shape like a conventional TIG power supply.

When producing this AC output, the frequency can be increased from the normal 60 cycles. Higher frequencies produce a narrower, deeper penetrating arc. The frequency is adjustable; about 125 cycles per second is often found to provide superior results on thinner materials.

Equal time intervals of DCSP and DCRP are also not necessary. To effectively clean aluminum, 15- to 25-percent DCRP is usually sufficient to provide the needed action. This avoids excess tungsten electrode heating and allows a longer DCSP penetration period. The percentage

Fig. 4.23. The last variable to set is post gas flow. It is important to protect the molten metal as it solidifies and cools at the end of the weld. The TIG torch is held stationary over the weld during the post flow. In addition to the weld pool, it is important to allow the tungsten electrode tip to cool from its 6,000 degree F temperature in a shielding gas atmosphere to prevent contamination.

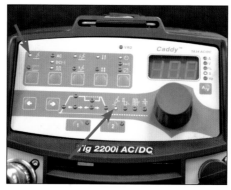

Fig. 4.24. The power source can also be set for DC and pulsed DC. Here, LEDs are lit for pulse (arrow). The LED for preflow is also lit (arrow), and the value shown is 10 seconds. The preflow time allows the shielding gas to purge the torch gas hose, the TIG cup, and most important, the weld start area of moisture-laden air.

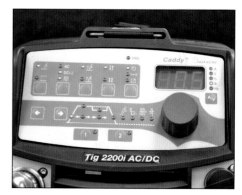

Fig. 4.25. Upslope time is set at 0.50 second because the pulse sequence continually varies the current as welding progresses. The use of pulse current provides an enhanced ability to control the weld puddle. This is particularly beneficial when welding sheet metal to control penetration and burn-through.

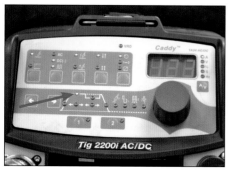

Fig. 4.26. The peak current is set at 123 amps. With pulsed TIG welding, to ensure good penetration, a higher peak current level can be set than is typically used for a given thickness material if pulse were not employed. Note the LED is lit for high current graphic (arrow). Turning the rotary knob on the right sets the value of peak current.

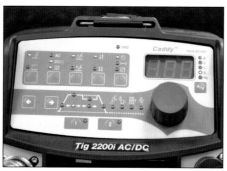

Fig. 4.27. Next, the amount of time spent at peak current is set. In this instance, 1.00 second is set. Therefore, the high current that could probably cause excess penetration is only on for a short time until the current is lowered. This avoids melting through, even with moderate manual manipulation skill.

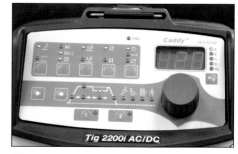

Fig. 4.28. A low-current level is set, which maintains some amount of arc heat but reduces the amount of penetration. The TIG torch is moved forward and paused along the weld seam at this low-current level, set here at 20 amps. It can also be used to pace adding filler metal. It helps to produce a uniform stack-of dimes weld appearance that some weldors try to achieve.

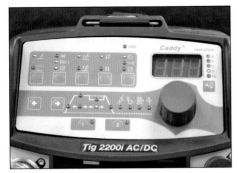

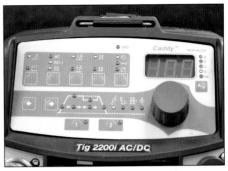

Fig. 4.29. The welder is set at the low-current level for a certain amount of time (1.11 seconds) and then it returns to the higher current level. Note the lit LED representing the time line. This light is lit in sequence by simply pushing the right arrow button (on the left). Pressing the left arrow button allows you to back up and adjust the previous parameter.

Fig. 4.30. Either a TIG torch switch, if using a sequential method of activating the welding steps, or a foot control activates the stopping and downslope time sequence (arrow). At this time, you can control the weld crater. It should be easier when using pulsed current because the puddle depression is reduced at the low-current cycle.

Fig. 4.31. Postflow of shielding gas protects the molten and cooling weld deposit from atmospheric contamination. For the 10 seconds, the torch should be held stationary above the weld deposit. It is also important to keep the gas flowing to eliminate contamination of the hot tungsten tip, preventing contamination and increasing its life.

of each, called balance, is usually adjustable.

With inverter-based TIG power, the DCSP and DCRP balance can be controlled over a wide range. Figure 4.15 shows a modern, inverter-based, microprocessor-controlled TIG welder that has a 220-amp capability and weighs only 33 pounds. An optional water cooler, for water-cooled TIG torches, mounts at the bottom and adds only 10 pounds to the system. Figure 4.17 shows the balance set for 76-percent DCSP and therefore 24 percent is DCRP. The greater amount of DCSP allows more weld penetration for a given amperage. Therefore, this inverter-based system can achieve the same amount of weld penetration at a lower current than conventional sine-wave AC TIG power.

It's very easy to set this type of microprocessor-based inverter TIG welder. A series of LEDs and a simple flow-chart-type display show which parameter is being set. Figures 4.16 through 4.31 show the welding parameters programmed into the welder. These include shielding gas preflow time, upslope time to reach peak current settings, the time at peak current, a low-current background current level, the time spent at the low-current level, slope time when stopping, and the post gas flow time when welding is stopped.

This particular unit has a number of other features, such as a foot-pedal option. It comes standard with a four-stroke control that allows welding without a foot pedal by simply energizing a button on the torch to activate the welding sequence. As mentioned earlier, this is like changing gears in a motorcycle, in which one movement sequences the operation. This kind of control is useful for welding tube intersections and welding overhead where a foot pedal is difficult to use or impractical.

The power supply also minimizes interference of high-voltage starts by limiting the amount of time it's used. If desired, a low-current-sensing, torch-work-starting circuit can be used instead of a high-voltage spike.

Fig. 4.32. If only DC welding is needed, a lower-cost DC-only version of the AC/DC inverter system is available, which limits the materials that can be welded to essentially steel and stainless steel. It has similar control features and is set the same way as the AC/DC power supply, but it's less complex and only weighs 21 pounds. The choice is based on the need to weld aluminum or magnesium. These materials require AC to enable some DCRP cleaning.

In addition, a special micro-pulse option is available for DC welding that minimizes welding heat on thin materials.

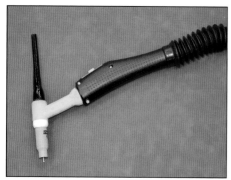

Fig. 4.33. Air-cooled torches are the most popular and are used at lower current levels for automotive-type materials. Water-cooled torches are more complicated to use and also more expensive but may be needed to access the weld joint for some weldments. A recently introduced, ergonomic handle version of the HW-17 called a TXH 200 is shown here.

An interesting feature is tungsten preheat. It is set as a percent of total current and helps with starting, and improves electrode life.

If only DC welding is needed, which limits the materials that can be welded to steel, stainless steel, and a few others, a lower-cost DC-only version of the AC/DC inverter system is available. It has similar control features and is set the same way, but it's less complex and weighs only 21 pounds.

Which power source you select is based on the need to weld aluminum or magnesium. These two materials require the use of AC to enable some DCRP cleaning of the weld area. If AC is not used, the high-temperature oxides that form prevent proper weld wetting.

TIG Torches

A variety of TIG torches are on the market today, and they fall into two basic categories: air cooled and water cooled. Both types have their areas of application. The choice is

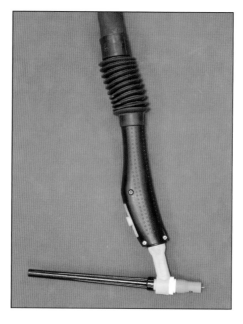

often based on a need for a small torch head for improved access to the weld joint, rather than absolute current level.

Air-Cooled Torches: These are the most popular and are typically used at the lower current levels used for automotive welding. These torches are the most useful without the cost and complication of water-cooling.

More than 18 models are available, and each has different-length cable assemblies. The most popular are the original Heliarc, HW-17, and HW-26. Other manufacturers identify their torches as Type 17 or Type 26. The HW-17 has a maximum capacity of 150 amps at 50 percent

Fig. 4.34. Several types of water-cooled torches are available. Those that use water to cool the power cable and the TIG torch head are smaller and lighter. Since circulating water cools it, the power cable contains less copper and therefore is lighter. An ergonomic version of the popular HW-20, called the TXH250, is shown here. It has a capacity of 250 amps, and it is significantly smaller and lighter than its 200-amp air-cooled counterpart.

duty cycle, and the HW-26 has 220-amp capacity.

Water-Cooled Torches: These TIG torches use circulated water to cool the power cable and the TIG torch head and, as a result, the torch can be smaller and lighter. Since circulating water cools it, the power cable contains less copper and is therefore lighter. If doing only a small amount of welding, you can get the water from a typical shop supply and the return water can simply be dumped in a sink drain. A professional approach is to use a circulating water cooler that includes a radiator and fan to dissipate the heat. Some manufacturers integrate the cooler into the welder. A satisfactory cooler could be constructed from a small car radiator or perhaps a large automatic transmission cooler employing an electric water pump.

Fig. 4.35. TIG welding is very sensitive to gas shielding quality. Unlike MIG welding, in which a small amount of air is tolerable while still producing a sufficient quality gas shield, TIG is very sensitive to any oxygen or nitrogen contamination. The gas lens was developed to improve gas shielding when TIG welding with even minor drafts of air.

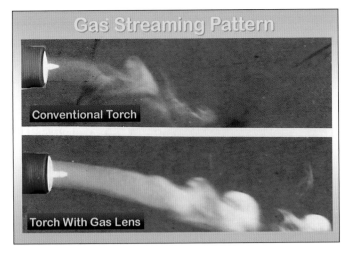

Gas Streaming Pattern

Conventional Torch

Torch With Gas Lens

Fig. 4.36. The quality of the shielding gas stream exiting a TIG torch with (bottom) and without a gas lens (top). Note how much longer the gas stream remains coherent as it leaves the torch with a gas lens.

For automotive-type welding, the compact, water-cooled HW-20 TIG torch is much smaller than the air-cooled HW-17 and allows access to intersecting tube joints. The HW-20 can also weld up to 250 to 300 amps while the heavier HW-17 only has 150-amp capacity.

An ergonomic version of the HW-20, called the TXH 250, uses all of the HW-20 accessories. Figure 4.34 shows the TXH 250 torch with a long cap, which holds full-length tungsten. However, for confined space welding, a very short flat-back cap is used. Short tungsten electrodes must be used when short back caps are employed.

Gas Lens: TIG welding is very sensitive to gas shielding quality. While MIG welding can tolerate some oxygen and nitrogen contamination, TIG welding and its tungsten electrode cannot. With MIG welding, a small amount of air is tolerable and still produces an acceptable-quality gas shield, but that's not the case with TIG. The TIG torch gas lens was developed to obtain improved-quality gas shielding when exposed to even minor drafts. Using a gas lens also allows the tungsten to extend farther out from the cup to obtain better weld visibility and joint access.

A conventional TIG torch and one equipped with a quality gas lens are shown in Figure 4.36. Note the columnar coherent gas stream is 4 to 5 times longer when the gas lens is employed. The gas lens is incorporated into the collet body, the part that holds the TIG collet and tungsten electrode. It is much larger in diameter and contains a number of fine screens that produce improved gas shielding. The threaded end accepts special gas lens TIG cups.

Fig. 4.37. The gas lens assembly (top) is incorporated in the collet body (bottom), which secures the collet and tungsten electrode in the torch. They both screw into the torch using the threads visible on their right side. The ceramic cups that fit them are different as noted by the different diameter of the cup threads visible on the left side of both parts.

The use of the very fine screens creates a quality gas-shielding stream, but these are vulnerable to damage if exposed to the external environment. To prevent that damage and produce a high-quality weld, a coarser and more durable screen is used as the last in the stack of multiple screens. The coarser screen protects the much finer mesh screens that provide the quality shielding.

Gas Cups: A high-impact number-6 gas-lens TIG cup is shown in Figure 4.38. TIG cups are manufactured from ceramic materials of two main types: aluminum oxide and aluminum silicate. Aluminum oxide is pink to light pink while aluminum silicate is often gray to white. Designated as high impact, the alumina cups have good electrical insulating properties and high-impact resistance. This is the most-popular-type cup in use, and the melting point of aluminum oxide is high, about 2,800 degrees F.

Made from a naturally occurring mineral called lava, aluminum silicate cups are sometimes referred to as lava cups, and are often called

Fig. 4.38. The shielding-gas cup (bottom) threads onto the gas lens TIG collet body (top). TIG cups are manufactured from aluminum oxide (pink) and aluminum silicate (gray). The alumina cups are often designated as high impact. The melting point of aluminum oxide is about 2,800 degrees F while the melting point of aluminum silicate, also called lava, is only about 2,100 degrees F.

GENUINE HELIARC 6

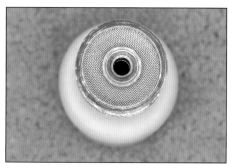

Fig. 4.39. A bottom view of the assembled gas lens/collet body as installed in the TIG gas cup. The lens covers the full opening. The tungsten electrode comes through the center hole. A major advantage of the gas lens is the ability to have the tungsten electrode stick out farther from the cup and retain quality shielding. Better visibility of the weld puddle is provided, so it's easier to watch melting and bead formation.

ceramic cups, although both types are ceramic. They provide good resistance to thermal shock at high temperatures and perform very well when high-arc heat is reflected back into the cup. However, the melting point of the aluminum silicate cup is about 2,100 degrees F and because of this lower melting point its maximum current rating is 250 amps.

TIG cups come in various outlet diameters and are designed to fit specific torch types. Large diameters are typically used for higher currents. They have internal threads that fit normal collet bodies and larger threaded inlet diameters to fit gas lens collet bodies. They must be selected to match the choice of torch and collet body.

Tungsten Electrodes

Several types of tungsten electrodes are used for TIG welding. The AWS has a specification and designation for the various types. To help

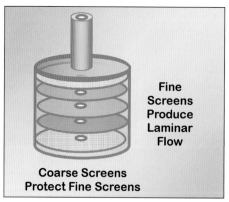

Fine Screens Produce Laminar Flow

Coarse Screens Protect Fine Screens

Fig. 4.40. Inside a quality gas lens collet body is a series of very fine screens, which cause the shielding gas to exit in a very smooth laminar pattern and stay coherent for a long length. One of the original patents explains that this produced significantly better performance than, for example, a porous bronze filter (see sidebar "TIG Shielding" on page 45 for more details).

identify each, a standardized color-coding at the electrode end is also defined.

Thorium Oxide: Pure tungsten is used for welding with sine-wave and conventional square-wave AC power because it has the ability to produce

a ball on the tip, which is needed for AC welding with this type of welding power supply. Tungsten electrodes with 1- and 2-percent thorium oxide were very popular for welding with DC power. When manufactured, the thorium oxide, or thoria, is finely dispersed in the tungsten, using powder metallurgy. These electrodes decrease the required energy for the electrons to leave the tungsten surface. As a result, more TIG energy is released in the plate, which is desired. However, thorium is slightly radioactive and has caused some concern.

Lantana: Other electrodes contain small amounts of rare earth elements, and therefore possess the same benefits as thoria. Electrodes are available with 1, 1½, and 2 percent lantana and can be used with similar performance as thoria, but lantana electrodes are not radioactive. With modern square-wave inverter power, they provide good performance with AC power. In addition, they provide excellent starting, good arc stability, and higher current capacity than pure tungsten electrodes. For sine-wave AC power, they can be conditioned

Tungsten Electrode Types			
Description	**Color Band**	**AWS Class**	**Weight % Range**
Pure Tungsten	Green	EWP	99.5% Tungsten (W)
1% Lanthana	Black	EWLa-1	0.8-1.2% La₂O₃
1 1/2% Lanthana	Gold	EWLa-1.5	1.3-1.7% La₂O₃
2% Lanthana	Blue	EWLa-2	1.8-2.2% La₂O₃
1% Thoria	Yellow	EWTh-1	0.8-1.2% ThO₂
2% Thoria	Red	EWTh-2	1.8-2.2% ThO₂
2% Ceria	Orange	EWCe-2	1.8-2.2% CeO₂
1% Zirconia	Brown	EWZr-1	0.80-1.20% ZrO₂
Rare Earth, etc	Gray	EWG	Not Specified

Fig. 4.41. This chart shows the types of TIG tungsten electrodes. Pure tungsten is used for welding with AC power using a sine wave and conventional square wave. Electrodes containing 1- and 2-percent thoria used to be very popular for welding with DC power. Electrodes are now available with 1-, 1½-, and 2-percent lanthana, which provide performance similar to thorium. With modern square-wave inverter power, lanthana electrodes can also be used on AC with good performance.

Tungsten Electrode Current Range and Cup Size (Using Argon Shielding Gas)				
			Balanced Wave AC [2]	
Diameter (inches)	Cup ID (inches)	DCSP [1] (DCEN)	Pure Tungsten	Rare Earth Added
.020	1/4	5-20 A	10-20 A	5-20 A
.040	3/8	15-80 A	20-30 A	20-60 A
1/16	3/8	70-150 A	30-80 A	60-120 A
3/32	1/2	150-250 A	60-130 A	100-180 A
1/8	1/2	250-400 A	100-180 A	160-250 A

(1) DCSP preferred. If DCRP used current range for 1/8 = 25-40 Amps
(2) If unbalanced wave AC reduced currents 30 to 50%

Fig. 4.42. Each electrode diameter has specific usable current ranges. When using DCSP power, the highest current range occurs where the tungsten heating is the lowest. When using DCRP, the maximum current reduces ten fold. AC power adds the high-electrode heating of DCRP and the maximum usable current must be reduced.

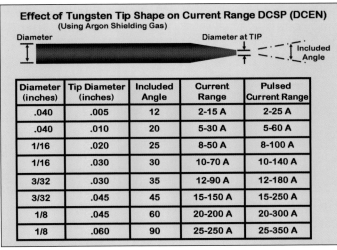

Effect of Tungsten Tip Shape on Current Range DCSP (DCEN)
(Using Argon Shielding Gas)

Diameter (inches)	Tip Diameter (inches)	Included Angle	Current Range	Pulsed Current Range
.040	.005	12	2-15 A	2-25 A
.040	.010	20	5-30 A	5-60 A
1/16	.020	25	8-50 A	8-100 A
1/16	.030	30	10-70 A	10-140 A
3/32	.030	35	12-90 A	12-180 A
3/32	.045	45	15-150 A	15-250 A
1/8	.045	60	20-200 A	20-300 A
1/8	.060	90	25-250 A	25-350 A

Fig. 4.43. The way the tungsten tip is prepared affects the maximum current-carrying capacity. This chart shows the significant variations in allowable maximum current with different tip angles and the size of the flat (tip) placed on the end.

with a balled end as required when using this type of power, which is not possible with thoria electrodes.

Lantana electrodes can operate with AC or DC power, and with fast switching inverter power they can be used with a pointed electrode, no need to ball the end.

Electrodes with zirconia and ceria additions are also available. Some electrodes use a combination of rare-earth arc stabilizing elements such as a mixture of zirconia, lanthana, and ceria.

However, for most applications, a 1½-percent-lantana electrode appears to provide the best overall performance and is a common recommendation, particularly with the more modern TIG inverter power supplies.

The useable current range of electrodes depends on the type of power, DC or AC, and if AC power the degree of unbalance. Figure 4.42 presents a guideline for acceptable maximum current for various-diameter electrodes.

Dressing: Preparing the tungsten electrode point can be done with a grinding wheel. The wheel is used solely for shaping the electrode point, so other materials are not transferred to the point causing contamination. Diamond wheels are best, but they

Fig. 4.45. The electrode should be ground to a uniform tapered point over a length that is approximately 2½ times the diameter. The grinding marks should be parallel to the electrode length. After the taper has been ground on the end, a very small flat is placed on the tip. This flat should be about .010 to .030 inch in diameter, depending on the electrode diameter. Be sure not to breathe any grinding dust.

Fig. 4.44. Bob Bitzky prepares the tungsten electrode point using a grinding wheel. The wheel should be dedicated to this use to avoid contamination. A fine mesh-grinding wheel should be used. Although diamond wheels are best, they are also very expensive. A good option is a 100-mesh silicon-carbide grinding wheel.

Fig. 4.46. Portable devices can be used to prepare the tungsten tip. They operate like a pencil sharpener, and employ a diamond disk wheel to provide an accurately sharpened point. The smooth surface they produce helps provide a very a stable arc. If performing a lot of TIG welding, it may be a desirable purchase.

are also expensive, so a fine mesh grinding wheel is the cheaper option. A 100-mesh silicon carbide grinding wheel is desirable.

The electrode should be ground to a uniform tapered point approximately over a length that is 2½ times the diameter. The grinding marks should be parallel to the electrode length. After the taper is ground on the end, a very small .010- to .030-inch-diameter flat is placed on the tip, depending on the electrode diameter. You need to be sure that the grinding wheel has a large protective screen so ground particles don't enter the face area. Be sure not to breathe any grinding dust, which is particularly important if using thoria electrodes.

Portable electrode sharpeners, similar to a pencil sharpener, are available, and these use a diamond disk wheel, which can provide a very accurately shaped point. The smooth surface they produce helps provide a very stable arc. If doing a lot of TIG welding, it may be a desirable purchase.

The shape of the end of the electrode is not only important in promoting arc stability, but it also helps define the current carrying capacity. Figure 4.43 provides the current carrying capacities of electrodes ground with different angles and tip diameters. For example, with a 3/32-inch-diameter electrode, the maximum current capacity can be increased 60 amps by using a 45-degree taper angle versus a 35-degree taper. Note the tip diameter (or end flat) is also increased from .30 to .45 inch.

Shielding Gas

Selecting the right shielding gas for TIG welding is easier than for MIG welding because there are fewer choices. TIG welding cannot tolerate even small amounts of oxygen or oxygen compounds, such as carbon dioxide. The shielding gas choices are argon, helium, or a mixture of the two. With the high price of helium, argon is usually the shielding gas of choice. With modern inverter TIG power sources, the ability to program various pulse parameters, and the improved wetting also makes helium less needed. This is especially true for the thinner materials associated with automotive welding.

Shielding gas flow rate is another factor to consider. Gas flow rate is measured in CFH. Note, don't confuse CFH with flow rate ratings of air com-

Selection of Shielding Gas		
Carbon Steel Chrome-Moly	Argon	Better arc starting and weld quality
	Spot welding; Argon preferred	Longer electrode life; better nugget shape
Aluminum & Magnesium	Argon	Better arc starting and cleaning action
	Argon-Helium	Hotter weld
Stainless Steel	Argon	Excellent control of penetration on gage material
	Argon-Helium	Hotter weld

Fig. 4.47. Choosing the shielding gas for TIG welding is easy because there are few options. TIG welding cannot tolerate even small amounts of oxygen or oxygen compounds, such as carbon dioxide. The shielding gas choices are argon, helium, or a mixture of the two gases. With the high price of helium, which is becoming a scarce commodity, argon is usually the shielding gas of choice.

Gas Flow Rates; Argon Shielding			
Cup Size ID	Typical Flow Rate	Max Flow Without Gas Lens	Max Flow With Gas Lens
# 4 (1/4 in)	8 CFH	15 CFH	~ 25 CFH
# 5 (5/16 in)	10 CFH	20 CFH	~ 30 CFH
# 6 (1/2 in)	12 CFH	25 CFH	~ 35 CFH
# 8 (1/2 in)	15 CFH	30 CFH	~ 45 CFH

Fig. 4.48. These are guidelines for usable TIG shielding gas flow rates. Note that excessive flow rates create turbulence causing air to mix into the gas stream. Using a gas lens is an advantage for TIG because it reduces the tendency for turbulence at higher flow rates. When using gas lenses, the suggested maximum flow rates should allow welding in up to a 4-mph draft. Beyond that, a windbreak of some type should be used.

TIG Shielding

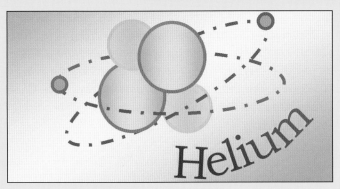

Fig. 4.49. Unlike argon, which is almost 1 percent of the earth's atmosphere, helium is only 0.0005 percent of the air we breathe. Helium is so light it usually requires more than twice the flow rate of argon. Helium prices have significantly increased over recent years, so look for alternatives when possible.

Helium is the second lightest element, next to hydrogen. While it is the second most abundant in the universe, it is only available in limited quantities on earth. Unlike argon, which makes up almost 1 percent of the earth's atmosphere, helium makes up only .0005 percent of the air we breathe. Where does all the helium go when a party balloon breaks or from TIG welding? Helium is so light it simply rises and leaves the earth's atmosphere. Helium gas is 10 times lighter than argon, which is 38 percent heavier than air. Argon helps in shielding the weld puddle. Helium is so much lighter than air that its use requires higher welding flow rates to achieve quality shielding. Flow rates are usually more than twice those of argon with helium or high-percentage helium mixtures.

Why is Helium so Expensive?

Although argon is extracted economically when manufacturing oxygen and nitrogen from liquefied air, this is not practical for helium. Most of the commercial helium comes from select natural gas deposits that contain sufficient percentages to make extraction practical. The United States has about half the world's reserves of these select natural gas deposits. However, the demand for helium is rising because of use in various industries.

Helium is used for medical devices, such as MRI body scanners, which use it for the superconducting magnets. The electronics industry also has a growing use for helium in crystal manufacture and other applications. Helium prices have significantly increased over recent years. Do not expect that situation to change. For welding purposes, look for alternatives when possible. ∎

pression usually quoted as cfm (cubic feet per minute). The Holley 850 in my 1934 street rod flows 850 cfm x 60 minutes per hour, or 51,000 cfh.

Another point of interest regarding gas flow rate is that we breathe at about 20 cfh of air with light activity, the same rate as typical TIG gas shielding flow. TIG shielding gas flow rate, not pressure, is set. The principle of choked flow is employed in flow-control devices to maintain the preset flow rate regardless of variations in flow restrictions that occur while welding. This becomes more important in MIG welding (see Chapter 6).

With small TIG cup sizes and lower currents, low flow rates can be used. TIG uses lower flow rates than MIG. However, if drafts are present during TIG welding, the low flow rates may cause air to be mixed with the shielding gas stream. Even small amounts of oxygen and nitrogen contaminate the tungsten and affect arc stability.

Excessive flow rate is also a concern because high flow rates create turbulence, causing air to mix into the gas stream. TIG welding is more sensitive to air mixing into the gas stream than MIG welding, and therefore the maximum flow rates suggested to avoid excess turbulence are lower. (Figure 4.48 presents typical and maximum suggested flow rates.)

TIG welding has certain advantages, and one of them is the ability to use a gas lens to reduce turbulence at higher flow rates. A gas lens should be able to accommodate a 4-mph draft using the maximum suggested flow rates; above 4 mph, a windbreak of some type should be employed.

Purchasing TIG Equipment

You need to purchase argon shielding gas. Check with your gas supplier and see what equipment it offers. Lower-cost alternatives are available.

For example, Harbor Freight offers lower-price equipment but you need to select the correct equipment for your needs. In addition, Sears has a wide selection of welders and several TIG welders from major welding equipment suppliers as well as several imported lower-cost units.

Comparison TIG Versus MIG Carbon Steel Weld Properties with ER70S-2 Rod/Wire					
Process	Shield Gas	Ultimate Strength (ksi)	Yield Strength (ksi)	% Elonga-tion	Impacts CVN @ −20 F
TIG	100% Ar	82	73	26	170 ft-lbs
MIG	98% Ar, 2% O$_2$	82	72	26	45 ft-lbs
MIG	CO$_2$	79	67	31	35 ft-lbs

Fig. 4.50. TIG produces weld metal deposits with very low oxide inclusions. Small oxide inclusions are created when oxygen enters the shielding gas stream due to turbulence or when carbon dioxide or oxygen is part of a TIG gas mixture. This chart compares all weld metal properties from a TIG weld made using an AWS ER70S-2 rod with a MIG weld made with ER70S-2 wire. The weld toughness, as measured by a Charpy test, shows the TIG weld is four times tougher. CVN stands for Charpy Vee Notch.

Look at all of the features on these alternate sources and find out how the particular unit will be repaired if necessary.

Selecting TIG Rod

TIG welding is very versatile and can join most materials. At first, it was used to weld aluminum and magnesium when no other effective process was available to weld those materials. It still is considered the process of choice to weld more exotic materials. Most fabrication is done with steel, and TIG welding is often used to weld this common material.

TIG welding is used to join a variety of metals and is particularly useful for more difficult to weld metals, such as aluminum and magnesium. TIG can use AC power and maintain a stable arc, and therefore, a portion of the cycle can be DCRP. DCRP provides the cleaning action needed on materials that form a high-temperature oxide, such as aluminum and magnesium. The use of AC power allows a portion of the weld cycle to achieve greater penetration and less tungsten heating with DCSP. The material being welded and the desired weld properties determines the correct welding rod to use.

Carbon Steel

TIG welding has a significant advantage over other arc welding processes, such as MIG welding, because the weld metal deposits can have very low oxide inclusions. When oxygen enters the shielding gas stream, small oxide inclusions are created in the weld deposit because of turbulence, or in MIG welding where it is mixed directly or is present in the form of carbon dioxide.

These small inclusions affect a parameter called toughness, and poor toughness allows a crack to move rapidly through a metal when it is quickly stressed. One measure of toughness is the Charpy test (see "Metallurgical Property Tests" on page 134). Suffice it to say, the energy required to quickly break a notched bar about 3/8 inch in cross section is often much higher when welding with TIG than MIG.

Figure 4.50 presents tensile and impact property data for a TIG weld made with an AWS ER70S-2 TIG rod compared to MIG welds made using wire of the same chemistry. As noted, the TIG weld takes about four times more energy to break than the MIG weld in the Charpy test. It takes 170 ft-lbs to break the TIG weld sample, and that's considered exceptional. The MIG weld made with 98-percent argon shielding gas required 46 ft-lbs to fracture, which is generally considered good crack resistance.

TIG-deposited weld metal can be very clean with low oxide levels, but the weld metal must start with a clean TIG rod, free from drawing lubricant, oil, and grease. Hydrogen, contained in these materials, contaminates a welding environment. Much more hydrogen can be absorbed in molten steel than steel at room temperature. Hydrogen can enter the weld when moisture-laden air is pulled into the shielding gas stream or from the surface contaminants on the welding rod.

AWS filler metal specifications do not define the amount of hydrogen

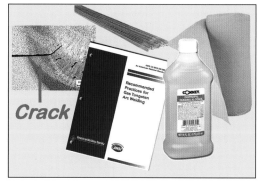

Crack

Fig. 4.51. Producing quality TIG weld metal requires a clean TIG rod, free from drawing lubricant, oil, and grease. Hydrogen is a very bad element in a welding environment because hydrogen from surface contaminants on the welding rod can enter the weld. An AWS publication states: "Prior to use, the TIG filler rod should be wipe tested with a solvent soaked cloth. After wiping, the cloth should show almost no sign of contamination."

compounds allowed on a TIG rod. Paraphrasing AWS A5.18, Specification for Carbon Steel Electrodes for Gas Shielding Arc Welding: "Rods must be free from foreign matter that would adversely affect the properties of the weld metal." A very informative AWS publication, *C5.5 Recommended Practices for Gas Tungsten Arc Welding*, contains a great deal of information about TIG welding. It states, "Prior to use the filler rod should be wipe tested with a clean cloth moistened with an approved solvent such as isopropyl alcohol. After wiping, the cloth should show almost no sign of contamination."

In my experience, the TIG rod may have some residual drawing lubricant or oil from the machine that straightens and cuts the rod. If it does not pass the recommended wipe test, it should be cleaned with a solvent, such as denatured alcohol. In addition, TIG rods should be stored in a sealable rod holder such as the one shown in Figure 4.52.

This is especially important when welding high-strength steels, such as 4130 chrome-moly, because

this relatively high-carbon steel (from a welding standpoint) is susceptible to hydrogen cracking. Even a small amount of hydrogen causes cracking in high-strength steels. (See Chapter 6 for more on carbon-steel filler metals.)

TIG welding is the preferred process for welding high-strength 4130 tubing used for race car chassis and roll cage construction.

4130 Chrome-Moly

Chrome-moly tubing has been used for airplane construction since World War II. Although considered weldable, it contains .33-percent carbon, which is much higher than

more readily weldable lower-carbon steels that contain about half that amount. In World War II, before TIG welding was in wide use, oxyacetylene welding was commonly used to join small-diameter 4130 tubes. This provides a very slow weld cooling rate and avoids production of a hard brittle material (martensite) that can form in 4130.

Most 4130 tubing is used in what is referred to as normalized condition. Normalized tubing has a tensile strength of about 95 ksi versus about 40 ksi for mild steel. NASCAR and most street rod chassis are fabricated from mild steel. The different strength levels for various heat treatments of

4130 Strength and Ductility with Various Processing

Material Treatment	Tensile Strength (ksi)	Yield Strength (ksi)	% Reduction in Area	% Elongation
Normalized 4130 Processed at ~1600 F Cooled slow enough to avoid Martensite	95	65	60	25
Annealed 4130 Processed at ~1600 F Cooled very slowly	82	53	56	28
Q & T 4130 Tempered at 400 F	230	205	40	10
Q & T 4130 Tempered at 1000 F	150	140	62	21
Q & T 4130 Tempered at 1200 F	115	100	64	22

Fig. 4.53. Normalized tubing is cooled moderately slowly from about 1,600 F and has a tensile strength of about 95 ksi versus about 40 ksi for mild steel. The different strength levels for various heat treatments of 4130 are shown in this chart. Note the wide range of properties depending on how the steel is processed.

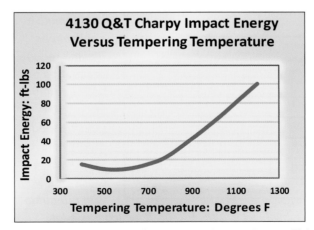

4130 Q&T Charpy Impact Energy Versus Tempering Temperature

Impact Energy: ft-lbs

Tempering Temperature: Degrees F

Fig. 4.54. Heat-treated 4130 tubing is quenched in water or oil from 1,675 degrees F. A very hard, brittle structure called martensite is formed; its strength is more than 225 ksi, but it is as brittle as glass. It is then reheated with a process called tempering, which lowers the strength and (depending on the temperature) improves the toughness. This graph shows a measure of toughness at various tempering temperatures, which is the result of a Charpy test.

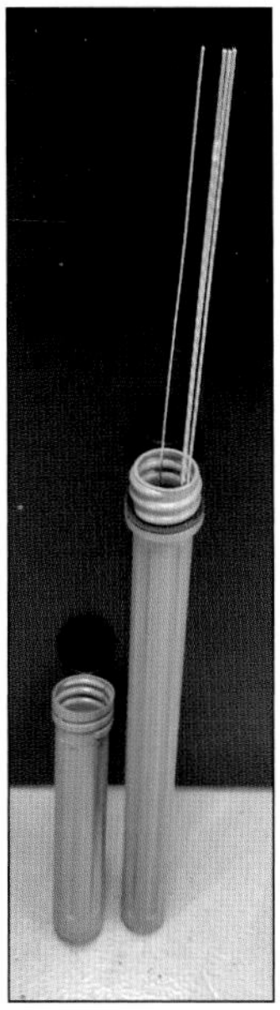

Fig. 4.52. A very clean TIG rod is especially important for welding high-strength steels, such as 4130 chrome-moly. Even a small amount of hydrogen causes cracking in high-strength steels. Store TIG rods in a sealable rod holder after a package of TIG rods has been opened and checked for cleanliness (and cleaned with a solvent if needed).

4130 are shown in Figure 4.53. Note the wide range of properties depending on how the steel is processed.

To manufacture normalized 4130, the tubing is cooled relatively slowly from about 1,675 degrees F, avoiding the formation of the high-strength, brittle martensitic structure created when rapidly cooled. The strength of normalized 4130 material with the high level of carbon incorporating chrome and moly alloy additions is still double that of mild steel. Normalized 4130 is the most common form of 4130 used for welded chassis and roll cages. It is often supplied as DOM, which refers to a processing method where it is drawn over a mandrill. DOM relates to the tubing straightness and sizing, but the metallurgical structure is normalized. If the tubing is quenched in water or oil from the 1,675-degree temperature, it forms the very hard martensitic structure and has a tensile strength of more than 225 ksi. Sounds great, but that high strength also makes the steel very brittle.

The optimal combination of strength and toughness is achieved by quenching to form martensite and reheating to lower the strength but improve the toughness. For example, if after quenching, the tubing is heated to 1,200 degrees F and the tensile strength decreases to 115 ksi, but toughness increases from being very brittle in the quenched condition to a level 10 times higher.

Figure 4.54 presents a graph of toughness compared to tempering temperature. Note there is a range of temperatures where the strength decreases but toughness does not improve and may even decrease. This range is 500 to 750 degrees F and must be avoided for structural applications.

Chrome-Moly Filler Rods: Do not consider welding with a filler rod that matches the chemistry of 4130 unless the finished welded part is going to be heat treated. If the part is to be heat treated, it is heated to 1,675 degrees F after welding, then water-quenched, and tempered. Therefore, a similar chemistry rod is used only if the part is heat treated. However, to avoid weld cracking, the material should be preheated to about 400 degrees F or higher and slowly cooled before heat-treating, to avoid cracks.

Most welding on thin 4130 used in automotive applications is done without preheat or heat treating after welding. Therefore, the main objective is to produce a crack-free weld with adequate but not excessive strength. For welding purposes, .33-percent carbon in 4130 is considered high, and it makes the weld deposit crack sensitive. Several AWS Specification alloy rods can be considered.

Many years ago, a fabricator of 4130 dragster chassis called me and asked for a TIG filler rod recommendation. At the time, I was managing a welding research laboratory for Linde, a leading supplier of filler metals and welding shielding gases. I took his desired welding conditions into consideration. He didn't want to preheat or use post-weld heat treatment. Because most joints were TIG fillet welds, a crack-free deposit was most important.

The recommended rod was the low-carbon content AWS ER70S-2. The weld deposit includes considerably more carbon and some chrome and moly from mixing with the melted tubing, increasing the weld deposit strength. The recommendation provided the high-quality welds desired and has become widely used when welding thin-wall 4130 tubing.

Some believe a rod containing moly must be used and suggest ER80SD-2. However, 4130 has .18-percent moly, while ER80SD-2 has .50, which is more than twice the amount. Also, another element, manganese, adds to the hardenability of the deposit. ER80SD-2 has more than three times the manganese as 4130. For fillet welds in 4130 tubing, a slightly larger fillet size allows ER70S-2 to provide the strength needed to match that of 4130 normalized tubing.

AWS Filler Metal Alloys Used to Weld 4130 Chrome-Moly Tubing			
	Best	Good	Fair	
Element	AWS ER70S-2	AWS ER70S-6	AWS ER80SD-2	4130 Tubing
Carbon	.04	.10	.10	.33
Manganese	1.15	1.65	1.85	.53
Silicon	.50	.86	.65	.05
Chrome	none	none	none	.90
Molybdenum	none	none	.50	.18

Fig. 4.55. Most 4130 welding in automotive applications doesn't require preheating or heat treating after welding. The main objective is to produce a crack-free weld with adequate but not excessive strength. The best rod is AWS ER70S-2, which has a low carbon and alloy content. The weld deposit includes considerably more carbon and some chrome and moly from mixing with the melted 4130.

Fusion Weld and 30% Dilution With Filler Metal			
Element	4130 No Filler	AWS 80S-D2	AWS 70S-2
Carbon	.33	.26	.24
Manganese	.53	.93	.72
Silicon	.05	.23	.18
Moly	.18	.28	.13
Chrome	.90	.63	.63
Structure	Martensite	Some Martensite	Little Martensite
Critical Diameter	2.4 inches	3.3 inches	2.1 inches
Crack Potential	High	High to Moderate	Moderate

Fig. 4.56. This chart shows the chemistry of welds made in 4130 chrome-moly with no filler metal and with 30-percent ER70S-2 and ER80SD-2 TIG rod diluted with 70-percent base material. Critical diameter is calculated, which defines the size of a bar that contains 50-percent martensite in the center after heating to 1,675 degrees F and quenching in water. The weld made with ER80SD-2 in this highly diluted weld has a significantly larger critical diameter than the other two.

Figure 4.56 shows what the chemistry is for a weld made in 4130 chrome-moly with: 1) no filler metal (called an autogenous weld) and 2) with 30-percent TIG rod and 70-percent base material using both ER70S-2 and ER80SD-2. Critical diameter is calculated for the resulting weld deposit. It defines the diameter of a bar that contains 50-percent martensite in the center after heating to 1,675 degrees F and quenched in water. (The method of calculating this parameter is defined in *Hardenability and Steel Selection* by Crafts and Lamont.)

The larger the diameter means the steel is more hardenable. Note the critical diameter for a weld made in 4130 without filler metal is 2.4 inches; for a diluted weld made with ER70S-2, it is slightly less at 2.1 inches; with ER80SD-2, in a diluted weld, it is 3.3 inches. This means the weld made with ER80S-D2 has a critical diameter 37-percent higher than a weld made by just melting the 4130. For a given welding condition and cooling rate, it has the possibil-

ity of being substantially harder than the base material.

An estimate of weld cracking susceptibility is also shown. The weld made without filler has high crack susceptibility, which is mainly due to the high carbon content. The weld with ER80SD-2 benefits from its lower carbon content but has higher hardenability, so the potential is rated high to moderate. A weld made using ER70S-2 is labeled moderate because it has the lowest carbon and its total alloy content makes the hardenability the lowest.

The weld strength may be somewhat lower than normalized 4130, but when fillet welding, a slightly larger fillet can compensate. In addition, an undermatch in strength is much better than a crack. Cracking is determined not only by weld chemistry but also by the amount of restraint the cooling weld is subjected to. Multi-tube intersections have high restraint and are more susceptible than a simple butt weld.

Some fabricators use ER80S-D2 rod, which may be desirable for

heavier section butt welds where less of the 4130 material is melted in the weld deposit. However, on these heavier sections, some preheat should be used to slow the cooling rate to avoid martensite in the deposit. ER80S-D2 is sometimes selected because the AWS designation for ER70S-2 indicates the minimum strength is 70 ksi. However, that is the minimum, not the typical, strength.

Figure 4.50 presents the results of an all weld metal TIG deposit made with ER70S-2 rod. It has a tensile strength of 82 ksi. That is a deposit without dilution into the high-carbon 4130. When using ER70S-2 rod to weld 4130 tubing, depending on the amount of filler rod and 4130 melted in the weld deposit, the strength will be higher and could approximate the strength of normalized 4130.

Aluminum

TIG is the ideal process for welding the aluminum alloys typically used in automotive applications where thin sheet-metal sections are often employed. The ability to control the heat and the amount of oxide cleaning are two important

Fig. 4.57. TIG is the ideal process for welding the thin aluminum typically used in automotive applications. The ability to control the heat and the amount of cleaning achievable with AC power are two important features. Here, a corner weld is made with the addition of filler metal.

The Metallurgy of Chrome-Moly 4130

Carbon steel has an interesting characteristic that occurs when it is heated above 1,675 degrees F. It changes atomic structure to be face-centered cubic, where the iron atoms rearrange to have one in each face of a basic cubic structure. If quenched in water or oil from that temperature, the structure of the atoms changes and forms a very highly stressed material. The higher the carbon content the more this stressed condition exists. Chrome-moly 4130 with its .33-percent carbon content is sufficient to produce very high strength in this quenched condition.

If cooled slowly, the atoms rearrange into a cube with an iron atom in the center of the cube. The resulting metallurgical structure produces a material with high strength that is resistant to brittle fracture or, said positively, is considered tough. However, if 4130 is quenched in water or oil from 1,675 degrees F, it forms martensite, which is a structure in which one side of the cube is longer and the other is shorter with an iron atom in the center. When this occurs, the important aspect is that the crystal lattice is highly stressed, and the presence of carbon further stresses the lattice. As a result, the ultimate tensile strength of 4130 increases from 90 ksi (if cooled more slowly from 1,675 degrees F) to more than 225 ksi (if water-quenched from 1,675 degrees F).

Unfortunately, the high-strength 4130 becomes a very brittle, glass-like structure. To make the hard, brittle martensite more ductile, the cooled quenched steel is heated, and the stressed structure relaxes to some degree, which is called tempering. How much the metal is tempered is dependent on how high the temperature is raised. The resulting material is called tempered martensite.

In general, the higher the tempering temperature, the less brittle and lower the strength. For structural tubing, the tempering temperature should typically be higher than about 1,000 degrees F.

The addition of chrome and molybdenum alloys to steel provides increased strength. The amount of strength increase depends on the amount of alloy added and how quickly the steel is cooled from a temperature of about 1,650 degrees F. The .90 chrome and .18 moly contained in 4130 are sufficient to provide strength more than five times that of mild steel when heated and quenched in water or oil.

However, a tempering temperature from about 500 to 700 degrees F should be avoided. This is called temper embrittlement. In that range the strength decreases and the toughness does not improve and may even decrease. Temper embrittlement can increase because of undesirable trace elements found in steel, such as phosphorous, sulfur, antimony, tin, and arsenic. The cost of the steel rises when these trace or tramp elements are removed.

Steels, such as those developed for submarine hull

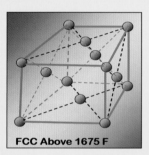

FCC Above 1675 F

Fig. 4.58. Carbon steel has an interesting characteristic that occurs when it is heated above 1,675 degrees F. It changes its atomic structure to being face-centered cubic (FCC), where the iron atoms rearrange to have one in each face of a basic cubic structure. If quenched in water or oil from that temperature, it changes the arrangement of the atoms and forms a very highly stressed material. The higher the carbon content, such as in 4130 chrome-moly, the more this stressed condition exists.

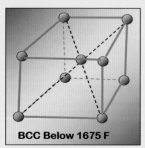

BCC Below 1675 F

Fig. 4.59. If 4130 chrome-moly is cooled more slowly, as occurs when producing a normalized condition, the atoms arrange into a cube with an iron atom in the center. This is called a body-centered cubic (BCC) arrangement. The slower cooling avoids producing the brittle martensite, and the resulting material is relatively strong and tough.

BCT Martensite Below ~600 F

Fig. 4.60. When 4130 chrome-moly is water quenched from 1,675 degrees F, it forms a highly stressed structure called martensite. The presence of carbon further stresses the lattice. This causes the 4130 to go from an ultimate tensile strength of 90 ksi, if cooled more slowly, to over 225 ksi if used as quenched. Unfortunately, a brittle structure accompanies this very high strength. It must be reheated, called tempering, to produce a practical structural material. BCT stands for body-centered tetragonal.

construction, can reach high strengths (130 to 150 ksi) and achieve very high toughness. These are also much easier to weld, having only .12-percent carbon. They also have very low undesirable residual elements but are higher in cost. These steels, unfortunately, are not readily available as structural tubing.

Figure 4.61 illustrates a continuous cooling diagram in green has been superimposed on a 4130 transformation diagram that uses a method defined by Grange and Kiefer. It shows the cooling rate needed to produce martensite. The cooling curves shown in blue and red dashed lines are cooling rates of TIG welds made in tubing thickness materials. These cooling rates form some martensite as an autogenous (no filler metal added) weld cools. The distance the green continuous cooling curve is moved to the right is referred to as hardenability.

For example, the larger critical diameter defined in Figure 4.56 for the weld made with ER80S-D2 places the continuous cooling curve farther to the right. For a given weld cooling rate, that deposit contains more brittle martensite.

Increasing stiffness is sometimes cited as the reason higher strength in desired. However, strength is a factor that does not directly determine stiffness. Figure 4.62 shows what occurs if a tube were welded to a stationary plate and a weight hung from the end. A carbon steel tube would bend an amount (l on the illustration) with a tensile strength of 40 ksi and a given weight (W) hung from the end of the tube. Assume the weight is selected, so it is not enough to cause the tube to take a permanent set (exceed its yield point). When the weight is removed, the tube returns to the original position. The distance (l) would be the same if that same weight is placed on the same size and thickness normalized 4130 tube or a heat-treated 4130 tube with 150-ksi tensile strength.

Modulus of elasticity is the property that defines the amount of bend, and all steels have essentially the same value. A stronger steel can handle a heavier weight before it takes a permanent set, but it just bends more as weight is added. The same-size tube made from aluminum weighs about one third that of steel, but bends three times the amount. A similar-size tube made from titanium weighs half that of steel, but it bends twice the amount. The modulus of elasticity is a function of the material and the values for aluminum and titanium happen to cause the amount of bend mentioned. ■

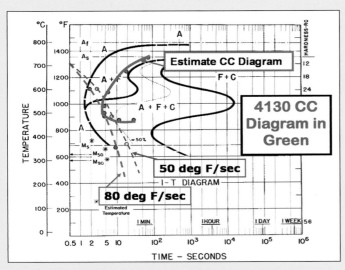

Fig. 4.61. The green line is the cooling rate needed to produce martensite in 4130 material. The blue and red dashed lines are cooling rates of TIG welds made in tubing thickness materials. These cooling rates form some martensite as the weld cools. The distance the green curve is moved to the right is referred to as hardenability of the material. Preheat slows the cooling rate. The black lines show transformation occurring at constant temperature.

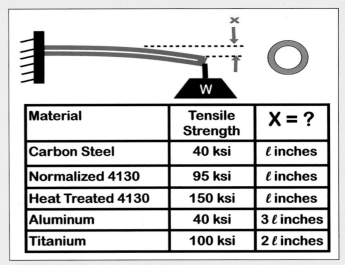

Material	Tensile Strength	X = ?
Carbon Steel	40 ksi	l inches
Normalized 4130	95 ksi	l inches
Heat Treated 4130	150 ksi	l inches
Aluminum	40 ksi	3 l inches
Titanium	100 ksi	2 l inches

Fig. 4.62. Increased stiffness is sometimes cited as the reason for using stronger steel. As shown here, any strength steel bends the same amount, if loaded with the same weight. This assumes the weight is not enough to cause the tube to exceed its yield point. Stronger steel handles a heavier weight before it takes a permanent set, but it bends more as the weight increases.

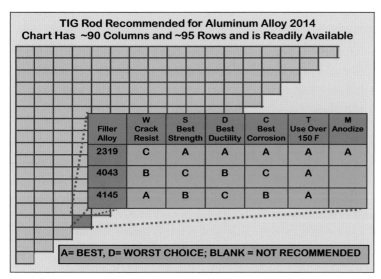

TIG Rod Recommended for Aluminum Alloy 2014
Chart Has ~90 Columns and ~95 Rows and is Readily Available

Filler Alloy	W Crack Resist	S Best Strength	D Best Ductility	C Best Corrosion	T Use Over 150 F	M Anodize
2319	C	A	A	A	A	A
4043	B	C	B	C	A	
4145	A	B	C	B	A	

A= BEST, D= WORST CHOICE; BLANK = NOT RECOMMENDED

Fig. 4.63. Selecting aluminum-welding rod from the many types available can be confusing. However, a table has been developed that makes the task easy, and a portion is shown here. Once the alloy to be welded has been defined, the rod choices are shown in one block in the table. In this instance, there are three acceptable alloys. There are six criteria shown with each block to help refine the selection.

features of TIG welding. AC power is used to weld aluminum where the DCRP portion of the cycle can be adjusted for the cleaning action needed. This power-source feature is called balance control.

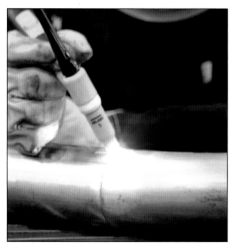

Fig. 4.64. TIG makes excellent welds in stainless steel. Much of the stainless in automotive use is thin sheet metal, and the ability to accurately control the heat with TIG makes it ideal for this application. Butt welds in stainless tubes for exhaust systems, as shown here, can be made with a minimum of filler metal.

There are many factors to consider when selecting the proper aluminum filler rod. A chart available from a number of sources (such as the *AWS Welding Handbook* and aluminum rod suppliers) makes it much easier to select the right rod for a particular job. Figure 4.63 presents a small section of the chart for joining two pieces of 2014. The green box detail indicates there are three rod alloys that could be used. The six criteria across the top help with the selection. Since these high-strength aluminum alloys are crack sensitive,

carefully consider the criteria for the three alloy choices.

Alloy 4145 is the most resistant to cracking and is labeled good for strength. However, if it is thin sheet metal in which restraint is low, alloy 2319 is best in all strength, ductility, and corrosion resistance. It produces weld strength of perhaps 45 to 50 ksi, significantly higher than the 2319, which may be about 30 ksi. However, it may crack in heavier sections.

Stainless Steel

TIG makes excellent welds in stainless steel because it provides accurate control of the heat input and therefore it can prevent warping. Much of the stainless steel that's welded in automotive applications is relatively thin, and heat control is critical for thin-gauge applications. The amount of filler added is also independent of the current employed. For example, butt welds in stainless tubes for headers can be made with a minimum of filler metal, using the necessary amount of current to make a fully penetrated weld deposit.

Using a stainless-steel TIG rod is less complex than with MIG. With MIG, shielding gas mixtures may contain some carbon dioxide, and

Typical TIG Stainless Steel Welding Wire Rods

Wire Alloy	Cr	Ni	Mn	Other	UTS, ksi	Yield, ksi
308*	20%	10%	2%		90	60
312**	30%	10%	1.8%	1.8% Mo	85	55
316 ***	19%	12%	1.6%	2.4% Mo	85	60

Note*   For welding 304 and most typical stainless base materials
Note**   Alloy 312 is often used for welding stainless to steel
Note*** Often used for marine applications. Moly addition helps with corrosion resistance.

Fig. 4.65. Type 308 TIG alloy rod is used to weld the most common stainless, type 304. If welding stainless to steel, a higher-chrome 312 alloy is often used. The other alloy that may be encountered is 316 stainless. It contains 2.4-percent moly to help with corrosion resistance and is often used in marine environments.

Fig. 4.66. Stiffness can increase with a larger-diameter tube. However, the weight increases with the diameter so a thinner wall must be used to maintain the same weight. Stiffness is proportional to the diameter cubed. Therefore, the stiffness of a tube twice the diameter is 8 times stiffer. A larger-diameter, thinner-wall tube can be used because it is both lighter and stiffer.

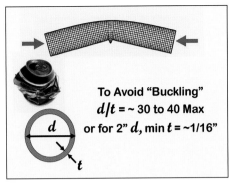

Fig. 4.67. There is a limit to how thin a tube wall can be before local buckling occurs. To avoid local buckling, a rule is that the tube diameter divided by wall thickness ratio should be less than about 30 to 40. This means that a 2-inch-diameter tube with a .093-inch wall thickness produces a ratio of 22, which is less than the 30 maximum suggested.

carbon pickup can cause corrosion problems. With TIG, pure argon shielding is used, so the there is no carbon coming from the gas, and therefore no corrosion.

Type 304 stainless steel is the most common and type 308 rod is used to weld it. A higher-chrome 312 alloy is often used for welding stainless steel to steel. Although some special names are used for this alloy that sound high tech, save money and use 312 instead.

Unless welding stainless to steel, do not use 312 rod for welding carbon steel or 4130. The welds look different but can cause cracking problems, depending on the ratio of base metal to rod in the deposit. If a large amount of base metal is melted and only a small amount of 312 rod is used, the desired stainless-steel deposit does not occur and a high-chrome steel weld can be very hard and the welds may crack. At best, it is a waste of money when joining carbon or 4130 steel.

Type 316 stainless steel is another alloy option. It contains an addition of 2.4-percent molybdenum and helps with corrosion resistance, and is often used in marine applications. That is more than 10 times more moly than the .2 percent in 4130 chrome-moly steel. As with most stainless steel used in automotive welding, the filler rod designation matches the plate alloy, except for type 304 stainless steel where type 308 rod is used.

Stiffness versus Diameter

Increasing the diameter of a tube can increase stiffness. That, however, also makes the tube heavier; the weight increases in direct proportion to the diameter. For a given wall thickness, the weight of a tube with twice the diameter is twice as heavy. The stiffness of a tube with twice the diameter is equal to 2 cubed, or eight times stiffer.

Therefore, a larger-diameter, thinner-wall tube can be used, because it is both lighter and stiffer.

As an example, switching from a 1.5-inch-diameter tube with a 0.187-inch wall to one with a 2-inch diameter and a 0.093-inch wall.

The weight ratio is calculated as:

$$(1.5 \times .187) \div (2 \times .093) = .66$$
Or:
$$1 - .66 = 34 \text{ percent less}$$

The stiffness ratio is calculated as:

$$2^3 \div 1.5^3 = 8 \div 3.4 = 2.4$$

Therefore, the tube would weigh 34-percent less and be 2.4-times stiffer.

There is, however, a limit to how thin a tube wall can be before local buckling occurs. This is sometimes referred to as the "crushed beer can effect." If a weight is placed directly on an empty beer can, the force needed to collapse the can is very high, much higher than possible to crush it by hand. However, if the beer can has a small dent placed on its side (for example, caused with a small amount of thumb pressure), the force required to crush the can is much lower. Try it!

In general, the tube diameter divided by wall thickness should be less than about 30 to 40 to avoid local buckling. (This rule can vary significantly depending on the type of loading and other factors.)

Continuing with the above example of a 2-inch-diameter tube with a .093-inch-thick wall, the buckling threshold can be calculated:

$$2 \div .093 = 22$$

The buckling threshold is 22, which is safely under the 30 guideline.

However, if the tube wall were only .062 inch thick, the ratio would be 32—within the risk zone.

Around 2005, NHRA allowed the use of heat-treated 4130 and higher design stress for some dragster rear frame tubes. There was a good deal of discussion and controversy about the decision at the time. In 2006 at a national drag racing event in Bristol, Tennessee, one of Don Schumacher Racing Team's dragsters crashed at about 300 mph. The crash was televised and the rear frame was seen to rapidly bend and break.

At the next televised drag race, TV announcers attributed the failure to local buckling. There is no official report on the cause of the failure or whether heat-treated 4130 tubing was involved. However, it appears from TV and other published comments that the tube wall may have been too thin. The TV announcers showed a sketch of a method employed in an attempt to prevent buckling from reoccurring.

Heat-Affected Zone

All the effort to provide increased strength by using quenched and tempered 4130 tubing is eliminated in an area of the tubing next to a weld. This is called the heat-affected zone (HAZ). Figure 4.69 shows a weld cross section with lines drawn to show this area. By definition, the weld was molten and reached about 2,500 degrees F. The base material directly next to the weld reached a temperature very close to 2,500 degrees F but did not melt.

Steel also has a transformation in structure that occurs at about 1,300 degrees F. (This structure change can be seen in a polished cross section with the proper etch.) The blue line shows the area that reached about

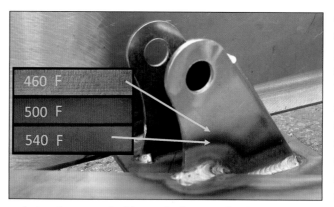

Fig. 4.68. This weld made in shot-blasted plate shows well-defined temper colors. There is a large distance from the weld edge to the area in the plate that reached 540 degrees F. With normalized 4130, the original metallurgical structure created is cooled more slowly and is not tempered martensite. Therefore, the HAZ in normalized 4130 has little strength reduction. The HAZ strength of heat-treated tubing could reduce significantly, depending on the original tempering temperature.

1,300 degrees F. Therefore, all of the work done to heat the tubing above 1,600 degrees F, water or oil quench, and reheat to temper (for example at 1,000 degrees F to obtain 150-ksi tensile strength) is eliminated in this area. The tempered martensite that was produced is now much higher in temperature and it is weaker.

Figure 4.68 shows a weld made in sand-blasted steel that has well-defined temper colors. These specific oxidation colors occur at certain temperatures. Note the significant distance from the weld to the area that reached 540 F. In fact, the area

that reaches 1,000 degrees F can be calculated if you know just two temperatures, such as the 2,500 and 1,300 degree F locations. At a minimum, this area is weaker than the carefully quenched and tempered original tubing.

It may also be possible that some brittle, tempered martensite is produced. With normalized 4130, the structure was cooled more slowly and martensite was never formed. Therefore, the HAZ in normalized 4130 has little if any martensite after welding. The strength might be slightly less in the HAZ but not lower than

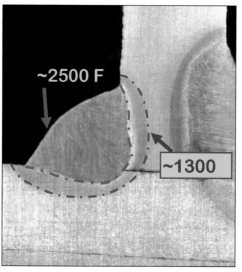

Fig. 4.69. About 2005, NHRA allowed the use of heat-treated 4130 and higher design stress for some dragster frame tubes. Use of heat-treated 4130 tubing has been suspected in several dragster and Funny Car chassis failures. Note the area next to the weld reaches temperatures well above any tempering temperature. In this heat-affected zone (HAZ), the strength of the tubing is reduced to levels of normalized tubing. If welding heat-treated 4130, the reduction in the HAZ strength must be considered in the structural design.

the annealed strength that is about 10 ksi less than normalized 4130 as seen in Figure 4.53.

Projects and Applications

Some application details should help provide an idea of where the TIG process is best applied. In some of the following examples the only good choice is TIG welding; in some examples other welding processes can be considered. The availability of TIG equipment and the proper welding skill are important considerations when making the process choice. In some instances, TIG may be slower but doesn't produce spatter, so welds can be made very flat, and there may be much less post-weld cleanup needed.

Project: Making a Taillight Bracket

A number of options were considered for taillights on this 1934 sedan pro/street rod. Modern lights did not seem to fit the classic Ford lines, and the original-style taillights did not fit the hot rod flames. Hence, it was decided to leave the body clean and make brackets to convert stainless steel 1934 Ford cowl lamp reproductions into taillights. The cowl lamps have odd-angle brackets and the stainless bumper brackets are also angled. A complex shape is needed to have the lenses flat and perpendicular to the road.

After laying out the angles, the best way to be sure the lenses are in the correct position is to make a cardboard mockup. It is a lot easier to trim cardboard than the .125-inch stainless from which the brackets are fabricated. (The assembled cardboard bracket is shown in Figure 4.72.) The next step is to cut the pieces with a hand plasma cutter.

The pieces are TIG welded, adding filler, and ground smooth when finished. The front, top, and bottom pieces are purposely cut slightly larger than needed so they could be ground to fit exactly after the back and two sides were joined. An angled top view of the finished bracket is shown in Figure 4.75. All welds are sanded smooth and polished.

1 *Fig. 4.70. To use 1934 Ford cowl lamp reproductions as taillights, fabricate brackets from 1/8-inch stainless steel. However, the cowl lamps have odd-angle brackets, and the stainless bumpers brackets are also angled. A complex shape is needed to mount the lenses perpendicular to the ground and car centerline.*

2 *Fig. 4.71. After laying out the angles, make a cardboard mockup to be sure the lenses are at the correct angle. It is a lot easier to trim cardboard than the .125-inch stainless from which the brackets were fabricated.*

3 *Fig. 4.72. Test-fit the assembled cardboard bracket and use scissors to make adjustments. The top, front, and back pieces were purposely made slightly larger than the model, so they can be trimmed during assembly. Next, cut the pieces with a plasma cutter.*

4 *Fig. 4.73. With a straightedge as a guide, use a plasma torch to manually cut these 1/8-inch 304 stainless-steel pieces. The edges of a plasma-cut surface have an oxidized surface. They also have some nitrides from the air used as the plasma gas. Grind this surface before welding to eliminate the possibility of nitrogen porosity and to eliminate any oxides.*

5 Fig. 4.74. Use a TIG welder with DCSP power and a 308 filler rod to weld the cut stainless-steel taillight parts. First, weld the back and two side pieces, being careful to match the angles of the cardboard model. Then use a grinder to trim the top, front, and back pieces, so they fit exactly before welding. Carefully sand the final welds smooth with progressively finer grit and polish the final assembly.

6 Fig. 4.75. This top view of this taillight assembly shows the compound angles required to mount it on the bumper bracket. The lights have the proper angles and are perpendicular to the ground. Note a braided stainless wire conduit is used to route the wires neatly under the fender where the electrical connections are hidden from view.

Application: Tubing Intersections

Fig. 4.76. This older chrome-moly dragster frame obviously failed during a crash. The weld is very small and broke with no deformation. With the proper size, shape, and quality of weld, the tubes should have deformed before this failure occurred.

Fig. 4.77. This modern roll-cage weld features a fillet weld that joins a rear brace to the top of a 4130 roll cage loop. For the size tube, it is small and very concave. This creates stress in the fillet throat as well as the thin areas at the edges of the weld.

A number of welds in street rods and race cars are tubular intersections. High-quality welds are necessary to handle the forces that might occur in a crash, which is why a roll bar is needed. Some believe that the smaller the weld the better, but that is not the case. Low-heat input is important, but TIG welds can be made too small for adequate structural support. In addition, some think concave welds are preferred, so they blend nicely into the mating tube, but they are prone to cracks and can have low overall structural strength.

Figure 4.76 shows an older chrome-moly dragster frame that obviously failed during a crash. The weld is very small. With the proper size, shape, and quality of weld, the

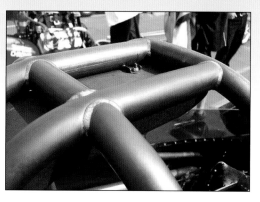

Fig. 4.78. The welds in this 4130 chrome-moly roll cage are large and slightly convex. They may not look as pretty, but form follows function, and these welds are structurally much better than concave fillets. (Photo Courtesy Jim Harvey, Harvey Racing Engines)

Fig. 4.79. This weld is joining 4130 chrome-moly tubing to a dragster roll cage. This weld has a size and shape appropriate for this size tube. It has less of a chance of cracking when the weld cools and should perform well, if needed, in a collision.

Fig. 4.80. The two welds joining 4130 chrome-moly tubes on this dragster frame are small and concave. They have only a small amount of welding rod in the weld joint. Depending on the type of filler rod, the large amount of 4130 in the weld nugget, and fast cooling rate, a weld this small could cause martensite to form in the weld, making it susceptible to cracking.

tubes should have deformed before this failure occurred. Figure 4.77 shows a more modern roll-cage weld joint that is also too concave. Figures 4.78 and 4.79 are welds in modern dragster roll cages, which are the proper size and are convex.

Figure 4.81 shows the stress problems with small concave welds. As the weld cools, it shrinks or attempts to shrink. The surface of the fillet weld is under tension in concave welds. The area near the weld edge is thin and faired into the other tube. When subjected to shrinkage stress, a crack could form. In fact, as stress and cyclical loading is applied, that crack opens. When a simple fillet weld cools between two plates, the vertical member is pulled toward the bottom member as the weld shrinks, which is often referred to as distortion.

In a tube-to-tube fillet weld, as the weld reaches the end circumference, the vertical tube is much stiffer than a flat plate. The cooling weld cannot pull the vertical tube significantly. Rather than contract, which it cannot do if highly restrained, the fillet and surrounding area are stressed just as if a load were being applied. This is referred to as residual stress.

Jim Harvey of Harvey Racing Engines supplied an informative series of photographs that illustrate the importance of fit-up required for making quality welds. These examples particularly apply to TIG welds in chrome-moly tubing. The weldor in Figure 4.82 is practicing safe welding. He is wearing flexible TIG welding gloves and a long-sleeved shirt. The arc is also well away from his helmet, keeping welding fumes from flowing into his breathing zone.

It's important to use a specially designed jig or drill press with a

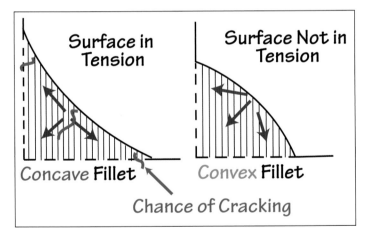

Fig. 4.81. Some weldors prefer the appearance of a concave weld, believing the faired-in edges are beneficial. However, small concave welds are prone to cracks and can have low structural strength. As the weld cools, the surface is placed in tension and the thin edges are subject to cracking. In a convex weld, the surface is not in tension. A flat to slightly convex fillet weld is the preferred shape.

Fig. 4.82. Harvey Racing Engines TIG welds a lot of 4130 chrome-moly tubing for race cars. Note the weldor is wearing flexible TIG welding gloves and a long-sleeved shirt. The arc is also well away from his helmet, keeping welding fumes from flowing into his breathing zone. (Photo Courtesy Jim Harvey, Harvey Racing Engines)

Fig. 4.83. Use a special-purpose jig or a drill press and a hole saw to bevel the ends at the correct angle. The contact surface must be flat for thinner wall tubing. For heavier wall thickness, a bevel is desirable so proper penetration is achieved. Here, an abrasive cartridge roll is used to finish the bevel surface on this tube joint. (Photo Courtesy Jim Harvey, Harvey Racing Engines)

Fig. 4.84. Properly prepared, there should be very little gap between the mating tubes. Regardless of the weld quality, these are considered partial-penetration welds and the unwelded notch creates a stress riser. The fatigue life of such a joint reduces the allowable stress significantly. (Photo Courtesy Jim Harvey, Harvey Racing Engines)

Fig. 4.85. This joint has considerable gaps and is not acceptable for critical weldments. The root defects are much larger than when fit-up is good. This leads to higher stresses in the root area and poorer fatigue performance. Slower weld speed and additional heat input may also be required. (Photo Courtesy Jim Harvey, Harvey Racing Engines)

quality metal hole saw to bevel the tube ends at the correct angles. The contact surface must be flat for thinner-wall tubing. For heavier-wall thickness, a bevel is desirable to achieve proper penetration. Structurally, these welds are considered partial penetration because the weld root is not fully fused. The non-fused

areas create a stress riser, and as the joint is cyclically stressed, a crack may propagate into the weld. This fatigue crack occurs when high loads vary, and loads certainly vary in a race car chassis.

In steel without defects, if the maximum stress is limited to half the ultimate strength of the material,

the fatigue life is considered infinite. (Note that is not the case with aluminum and the reason aluminum connecting rods can only be used for a limited time.) However, notches and cracks require reductions in the allowable stress to achieve infinite fatigue life. For partial penetration fillet welds, some structural design

Fig. 4.86. This tee weld has excellent fit-up. With this type of quality fit, it is relatively easy to make a sound well-penetrated TIG fillet weld. Tee welds are common joints and can be cut with a drill press and quality hole saw. (Photo Courtesy Jim Harvey, Harvey Racing Engines)

Fig. 4.87. This is a TIG weld made in a properly fit tee joint. It has a properly sized flat weld for the tube size. This fillet shape is preferred over a small concave fillet and is less susceptible to cracking. It also provides superior strength. (Photo Courtesy Jim Harvey, Harvey Racing Engines)

standards may require the allowable stress be reduced by 70 percent or more, but this depends on the amount of penetration, the presence of defects, etc. Poorly fitting joints create more and larger root defects.

In Figure 4.83, an abrasive cartridge roll is used to finish the beveled surface. The proper fit-up is shown in Figure 4.84. The joint shown in Figure 4.85, might at first appear satisfactory, but a careful look shows excess gaps exist. This causes slower speeds to be used and larger internal root defects.

Application: Stainless-Steel Gas Tanks

One of the specialties of Rock Valley Antiques Auto Parts is welding stainless steel for gas tanks. The staff made the special stainless-steel tank for a 1934 pro/street rod. Although the tank has some internal baffling, extensive baffling is used for cars with built-in fuel pumps and return lines.

Figures 4.90 and 4.91 show an Aeromotive A1000 high-capacity fuel pump mounted in a stainless tank. The pump flows 500 pounds of fuel per hour and can deliver at 90 psi. The tank has a 3/8-inch treaded outlet and a 1/2-inch return fitting. Stainless-steel clips are TIG spot-welded to secure the baffles, and a spiral baffle is attached to the top of the tank. It's used to keep fuel near the pump when the car is under heavy g loads, and those g loads push the fuel away from the pump. The oxidation from the TIG spot welds is seen in the uniform circular pattern surrounding the black pump cover.

One method of making a circular baffle is to cut triangular wedges from a flat plate and then make 90-degree beds at the point of the V. These bent shapes can be TIG tack welded to the top of the tank.

1 Fig. 4.88. Rock Valley Antiques Auto Parts fabricated the special stainless-steel gas tank to fit the narrowed rear frame on this 1934 pro/street rod. Although the tank has some internal baffling, units designed for cars that employ built-in fuel pumps and return lines have more extensive baffling.

2 Fig. 4.89. This is a view of the gas tank fit into the car. The flanges on the side were bolted into the narrowed frame rails on the pro/street chassis. It has several internal baffles to keep gas from moving rapidly while cornering or accelerating.

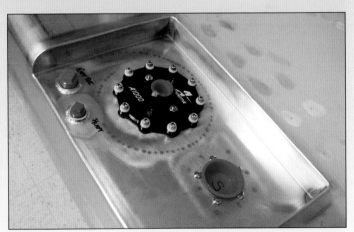

3 Fig. 4.90. Rock Valley mounted an Aeromotive A1000 high-capacity fuel pump into a custom stainless gas tank. The pump flows 500 pounds of fuel per hour and delivers at 90 psi, making it compatible with a fuel-injection system. (Photo Courtesy Dale Mathison, Rock Valley Antique Auto Parts)

4 Fig. 4.91. The stainless tank has a 3/8-inch threaded outlet and a 1/2-inch return fitting. Rock Valley TIG spot-welded the clips that hold the internal baffles. To ensure fuel remains near the pump, a spiral baffle is employed and attached to the tank top. The oxidation from the TIG spot welds are seen with the uniform circles surrounding the black pump cover. (Photo Courtesy Dale Mathison, Rock Valley Antique Auto Parts)

5 Fig. 4.92. In this design, a number of internal baffles prevents fuel from sloshing around under acceleration and when making hard turns. The bottom corners of the baffles are cut at an angle so there is enough space for fuel to move slowly from one section to another. (Photo Courtesy Dale Mathison, Rock Valley Antique Auto Parts)

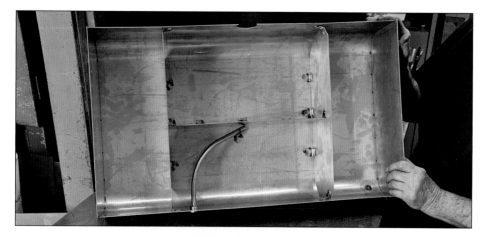

6 Fig. 4.93. The small stainless-steel clips are TIG spot-welded to the baffle and then TIG spot-welded to the tank bottom and sides. Several spot welds on each clip ensure they do not come loose. (Photo Courtesy Dale Mathison, Rock Valley Antique Auto Parts)

7 Fig. 4.94. This close-up shows how five spot welds join each clip. The multiple spots ensure the quality of the attachment of the baffles, which are subjected to loads from the fuel as it tries to move from one part of the tank to another as high g-load forces are applied. (Photo Courtesy Dale Mathison, Rock Valley Antique Auto Parts)

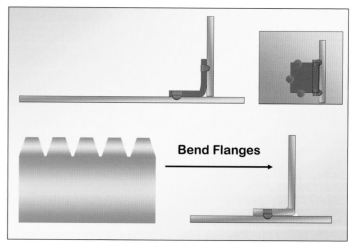

Bend Flanges

8 Fig. 4.95. This schematic shows two types of clips used to secure the baffles. The top one is visible in Figures 4.93 and 4.94. The bottom example is a way to make the curved baffle.

Project: Making a Stainless Exhaust System

A Borla hot rod kit is used to make the exhaust for the Chevy 502/502 in the pro/street rod. The kit comes with various mandrel bends and straight sections of 3-inch stainless tubing and straight-through stainless mufflers. The pieces are cut to fit the clearance available from the Sanderson long-tube headers to the outlets. Since there is no room in the pro/street chassis to go over the narrowed Ford 9-inch rear end and exit out the back, the exhaust was routed to exit just ahead of the rear wheels but behind the door openings. Pieces are cut, fit, and MIG tack welded.

The butt joints of the type 304 stainless tubes are TIG welded using the 308 rod. The exhaust pipes slip into the muffler openings.

Rather than rely on clamps, a MIG fillet weld is made after fitting and tack welding on the car. The MIG welds were made with our usual steel shielding gas mixture of 8-percent carbon dioxide, 2-percent oxygen, and 90-percent argon, which is unlike the pure argon shielding gas that is used for the TIG welds. Type 308LSi wire is used, which is a low-carbon wire with silicon added for better MIG weld wetting. The low carbon content helps prevent corrosion problems from the 8-percent carbon dioxide in the shielding gas. Do not use more than 10-percent carbon dioxide if MIG welding stainless steel.

1 Fig. 4.96. A Borla hot rod kit was used to make the exhaust for the Chevy 502/502 in our 1934 pro/street rod. The kit includes mandrel bends and straight sections of 3-inch stainless tubing and straight-through stainless-steel mufflers. Cut the pieces to fit the clearance available from the Sanderson headers to the outlets.

2 Fig. 4.97. Cut, test fit, and MIG-tack-weld the pieces. Bring the tacked sections and TIG welded the butt joints. Weld the type 304 stainless tubing with 308 rod.

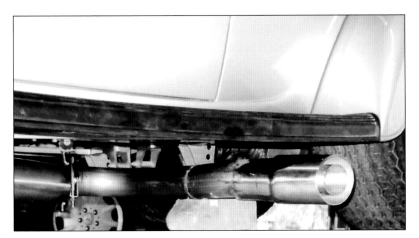

3 Fig. 4.98. Slip the 3-inch exhaust pipes into the muffler openings and lap fillet weld with MIG. Make the MIG welds with normal steel gas mix, which is 8-percent carbon dioxide, 2-percent oxygen, and 90-percent argon. Use type 308Lsi wire, which has a low-carbon content for improved corrosion resistance. Add silicon for better MIG weld wetting.

4 *Fig. 4.99. There is no room to have these large-diameter exhaust pipes pass over the tight space in the pro/street narrowed rear chassis. The exhaust exits in front of the rear wheels. The Borla exhaust tip is a dual-pipe system, keeping the reflected heat from the running board. MIG weld it after final cutting and tack welding on the car.*

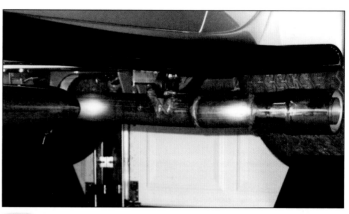

5 *Fig. 4.100. The TIG welded butt joints and exhaust tip are visible in this photo. The concentric pipe exhaust-tip system with its outer 4-inch-diameter cover pipe over the 3-inch exhaust pipe helps protect the running board rubber cover and prevents inadvertent contact.*

Application: Exhaust System

Fig. 4.101. Legacy Innovations, Inc., is a fabricator of race cars and pro/street rod show cars. Twin turbos sit high in the engine bay and the TIG-welded stainless intake and exhaust pipes are an artistic assembly of tubes.

Fig. 4.102. The stainless exhaust system on Legacy's pro/street S-10 is a remarkable sight. It is in the process of being fabricated in this view and is partially welded. It has many butt welds and slip joint fillet welds.

Legacy Innovations in York, Pennsylvania, is a very interesting race car and pro/street rod show car fabricator. The company's cars have appeared at SEMA and other national shows. The stainless exhaust system on its pro/street S-10 is a remarkable sight. Twin turbos sit high in the engine bay, and the stainless-steel intake is an artistic assembly of tubes. The stainless exhaust has many butt welds and slip-joint lap fillet welds. Considering the loads, and the fact that the butt welds are full-penetration TIG welds, these welds can be small. Most of the welds on the custom S-10 are made with TIG.

Legacy Innovations also has a Chevelle in the shop with a beautiful pair of stainless headers fabricated for the big-block engine. Heavy flanges ensure a leak-free fit.

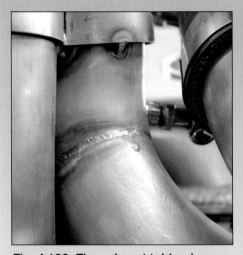

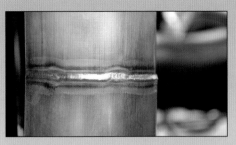

Fig. 4.104. This is an example of a well-fabricated butt weld. It has 100-percent penetration and a minimum of weld reinforcement. A properly welded, 100-percent penetration butt joint has a strength matching the stainless tubing.

Fig. 4.103. The exhaust tubing is carefully cut to provide gap-free butt joints. Small butt welds are fine for these exhaust-tubing welds. They are full-penetration welds and require a minimum of filler metal.

Fig. 4.105. The headers made for this pro/ street Chevy are a work of art. They are of equal length starting at the Chevy big-block motor. To achieve the proper lengths for each tube, the 2-inch stainless exhaust tubes required a number of bends.

Fig. 4.106. The right-side headers are equally impressive. One of the TIG butt welds is visible on a rear exhaust tube. The narrow HAZ demonstrates the low-heat-input of the TIG weld. It is flat and uniform in appearance. Since it is fully penetrated, it provides matching strength to the stainless tubes.

Fig. 4.107. This close-up of the exhaust system shows the heavy thickness of the flange. This provides a leak-free seal and avoids distortion in this harsh environment. The welds between the heavy flange and the thin tube are more difficult to make, but these are very well crafted. The uniformity of the HAZ attests to their quality.

Project: Repairing Aluminum Cylinder Heads

Aluminum cylinder heads eventually lose some of their heat treatment, and with the repetitive pounding of the valves, the hardened valve seat insert loosens and causes a failure. Assuming the failed seat does not create a catastrophic head failure, the valve seat area can be weld repaired. The example in Figure 4.111 is an old NASCAR cylinder head in which the valve seat area needs repair. The first step, as with

all aluminum welding, is to carefully clean the area of all contaminants, especially combustion carbon. The use of an aluminum chemical cleaner is also helpful and should be used for this critical application. Remove all visible carbon or oils then use carbide burrs to provide a smooth area to be filled with weld metal.

The ports in a high-RPM NASCAR engine are large, and there is not much metal left in the exhaust or

intake port area. An inverter square-wave power source set at 76-percent DCSP balance provides the needed penetration. This 220-amp system is set at the maximum current valve and uses a 250-amp water-cooled TIG torch. Preheat is needed to ensure the proper wetting and to prevent cracking as the weld cools.

Typically, 300 to 350 degrees F should be sufficient for most head configurations. Some recommend

Fig. 4.108. TIG welding is preferred for repairing aluminum cylinder heads because the amount of heat can be controlled independently from the amount of fillet rod added. Considering the high heat to which the cylinder heads are exposed, and the pounding to which the valve seats are subjected, it is no wonder a failure is inevitable. This is an old NASCAR head needing seat repair.

an even higher preheat, but 300 degrees F worked fine for this weld. A simple oven for cylinder head preheating is a propane-fired barbeque with a cover. Keep the flame on low while welding and when finished, put the head back on the grill, shut off the gas, and let it cool slowly.

Most aluminum cylinder heads are made from castings. Some are made from 355, 356, or 357 alloy aluminum, which contain about 5- to 8-percent silicon.

Type 4043, a common 5 percent silicon alloy welding rod, is used for this project and made sound well-wet welds. Type 4145, a 10-percent silicon alloy, is an alternate and one recommend by AlcoTec, a leading aluminum welding rod manufacturer. Type 4145 provides low susceptibility to weld cracking when used with aluminum-copper and aluminum-copper-silicon castings. AlcoTec, a leading aluminum welding rod manufacturer. Type 4145 provides low susceptibility to weld cracking when used with aluminum-copper and aluminum-copper-silicon castings. AlcoTec recommends it for repairing cylinder heads because it is useful for elevated temperature service.

Fig. 4.109. The area needing weld build-up must be very clean before welding takes place—any combustion carbon must be removed. Oil or grease must also be eliminated. An aluminum chemical cleaner is well suited for this critical application.

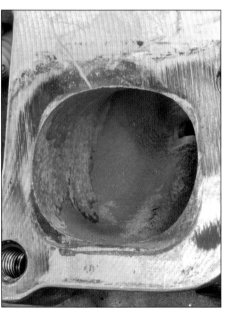

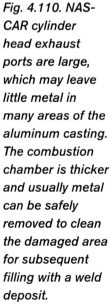

Fig. 4.110. NASCAR cylinder head exhaust ports are large, which may leave little metal in many areas of the aluminum casting. The combustion chamber is thicker and usually metal can be safely removed to clean the damaged area for subsequent filling with a weld deposit.

Fig. 4.111. The cylinder-head intake ports are very large, and care must be taken when grinding in areas around them. If an intake valve seat needs repair, caution is needed when removing damaged material. In this cylinder head repair, the exhaust valve seat is damaged and needed to be cleaned and built-up with weld.

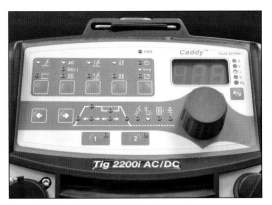

Fig. 4.112. An AC TIG welder made the weld repair on a cylinder head. Pure argon shielding gas is utilized. With very good cleaning of the area prior to welding, a 24-percent reverse-polarity cleaning cycle has been selected. This allows 76-percent straight-polarity for more penetration and heat in the cylinder head.

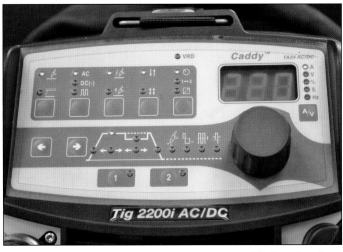

Fig. 4.113. Set the current at 220 amps, sufficient for the weld wetting and penetration needed. Preheat the cylinder head to 300 degrees F prior to welding. Preheat helps with wetting and reduces the possibility of cracking as the weld cools.

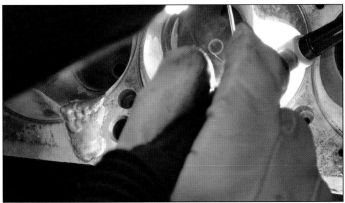

Fig. 4.114. Use a 250-amp water-cooled torch, model TXH 250W, to make the welds. The small torch head helped access to the joint. The overall welding time is short, so there is no need to for extra heating while welding.

Fig. 4.115. Use a 3/32-inch-diameter 4043 silicon-alloyed aluminum rod to make the weld deposit. Most cylinder heads are made from a silicon-alloy casting. Silicon reduces shrinkage stresses when cooling. An alternative with even a higher silicon content can be used, such as a 4145 rod. It is satisfactory for high-temperature service, such as cylinder head repair.

Fig. 4.116. Allow the finished weldment to cool slowly. There is sufficient weld metal to allow the subsequent machining operation to produce the required valve seat pocket. Note that the square-wave inverter-based power source and the 76-percent balance setting is allowed when welding with 220 amps. With a conventional AC TIG power source, more than 300 amps are required to achieve the same result.

Application: TIG Welding Titanium

Titanium is a difficult material to weld because of the potential for contamination from oxygen, nitrogen, and hydrogen. Even heated to 800 degrees F, it reacts with these gases and becomes brittle. Therefore, an excellent gas shield must protect the molten weld puddle. Until cooled below 800 degrees F minimum, the weld and HAZ must be protected. In general, argon is the preferred shielding gas because of its high density and shielding quality versus helium. Adequate shielding can be accomplished with very careful gas control using a large TIG gas cup with a gas lens and the addition of a trailing shield. Trailing shields are often custom made to fit the application.

In addition, the back of the weld joint must be protected with another argon line and appropriate device to ensure argon floods that area. It's necessary for argon to purge the air inside the tubing while welding. DCSP is typically employed when welding titanium.

Shops that specialize in welding titanium parts often use a sealed glove box, which can be purchased in various sizes or fabricated. No gas pressure is required, just a sealed chamber with a check valve outlet, so argon can purge air from the chamber before welding. The easy way to verify that all contaminants have been purged is to use a scrap piece of titanium and strike an arc on the surface, creating a small puddle. Stop welding and hold the torch stationary over the surface for about half a minute. The weld should be shiny with perhaps a slight straw color. Any blue, even light blue, indicates contamination.

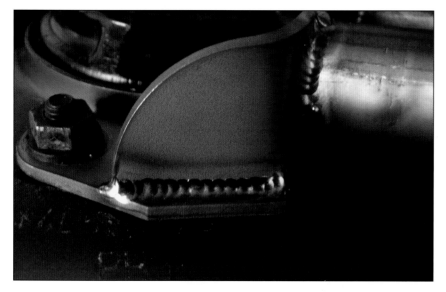

Fig. 4.117. Here is a titanium spring tower and strut brace. To avoid any oxidation, it was welded in a large, clear plastic bag. To complete the welding job, a pair of welding gloves and a TIG torch were taped to holes cut in the bag. Before welding started, the argon shielding filled the bag and purged all air. (Photo Courtesy Jim Harvey, Harvey Racing Engines)

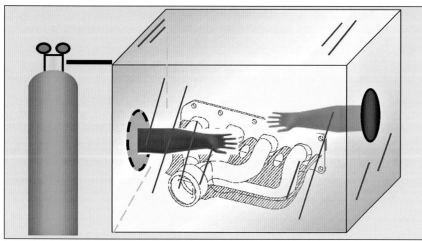

Fig. 4.118. If titanium is exposed to air down to 800 degrees F, it becomes brittle. If doing a lot of titanium welding, using a glove box with airtight seals is the easiest way to achieve the required shielding. Argon purges the chamber before welding. To check the quality of the purge, an arc is struck on a piece titanium scrap. The resulting weld should be bright and shiny and never a blue color.

Jim Harvey of Harvey Racing Engines welded the titanium strut tower brace shown in Figure 4.117. A large, clear-plastic bag with gloves taped into the openings was used to weld the end tower brackets to the tubular cross brace. The TIG torch cable was also taped through an opening in the bag. Gas flowed from the TIG torch until the bag inflated and argon gas escaped through a small hole. Sufficient time was allowed to ensure a quality purge of all air. Since Harvey Racing does not weld a lot of titanium, this approach achieved good shielding with minimum cost.

The Funny Car shown in Figure 4.119 uses a titanium bell housing. The high-quality welds on the housing stiffeners were probably made in a chamber or with a trailing shield and back purge. The car owner tack welded the bracket shown in Figure 4.121 in the proper position. The tacked assembly was then sent back to the manufacturer for final welding.

Fig. 4.119. Titanium can be as strong as chrome-moly and is about 40-percent lighter. This Funny Car uses a titanium alloy bell housing to reduce weight. The welds on the housing stiffeners are bright and free from contamination. (Photo Courtesy John Bray)

Fig. 4.120. The high-quality welds on the housing stiffeners were probably made in a chamber or with a trailing shield. If welded out of a chamber, a purge of the back side is also needed. (Photo Courtesy John Bray)

Fig. 4.121. The car owner needed a custom bracket welded to the housing. He carefully TIG tack welded the titanium bracket where it was needed, and then sent the assembly back to the bellhousing manufacturer to have it welded properly. (Photo Courtesy John Bray)

Application: TIG Welding Magnesium

Magnesium is 35-percent lighter than aluminum and has a similar low melting point of 1,200 degrees F. Like aluminum, it rapidly creates a very high melting point oxide that prevents weld bead wetting. Excellent shielding is needed to achieve sound, well-wet deposits. Unlike aluminum, it is difficult to bend at room temperature and needs to be heated to 400 to 600 degrees F to achieve good workability.

The two beads on the right in Figure 4.122 were made on magnesium sheet with argon shielding and AC power using a balance of 75-percent DCSP and 25-percent DCRP. The magnesium plate was purposely left with some surface oxide to see how well the AC TIG arc with a 75-percent DCSP balance cleans the surface. The aluminum TIG weld was made on a clean surface. On the aluminum, the cleaning band is very obvious and uniform on both sides of the weld bead. For the magnesium welds, the cleaning area is visible but is not as uniform because the surface is somewhat oxidized compared to the clean aluminum plate. The welds are sound and without porosity. Welding both aluminum and magnesium should always start with a very clean surface.

A corner weld made in 1/8-inch-thick magnesium sheet is shown in Figure 4.124. It was welded at 130 amps AC with a balance of 65-percent DCSP and 35-percent DCRP using argon shielding gas. The DCSP/DCRP balance was altered from the test welds to see if additional cleaning action could be achieved and to reduce the chance of burn-through. The cleaning action was about the same as with 25-percent DCRP. This autogenous weld (a weld made without filler metal addition) wet well and was porosity free.

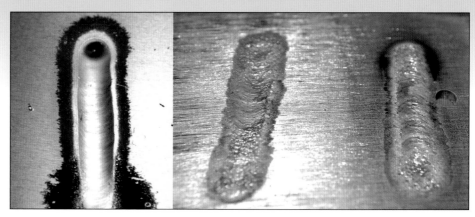

Fig. 4.122. Magnesium is 35-percent lighter than aluminum and has a similar low melting point of 1,200 degrees F. Like aluminum, it also oxidizes rapidly and therefore excellent shielding is needed to achieve sound well-wet deposits. The two beads on the right were welded with AC power using a balance of 75-percent DCSP and 25-percent DCRP. On the left is a similar-size weld made in aluminum.

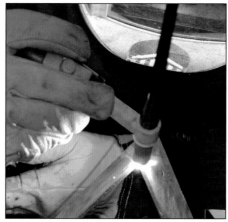

Fig. 4.123. This corner TIG weld is made in 1/8-inch-thick magnesium at 130-amps AC with 65-percent DCSP using argon shielding gas. The DCSP/DCRP balance was altered from the test welds to see if additional cleaning action could be achieved and to reduce the chance of burn-through. The cleaning action was about the same. This autogenous weld (meaning no filler added) wet very well and is porosity free.

Fig. 4.124. Here, a small amount of cleaning action is visible. The weld is sound and did not burn through. However, for this autogenous weld, avoiding burn-through would have been easier with somewhat lower power or a lower percentage of DCSP.

STICK WELDING

Developed early in the 1900s, stick welding is one of the oldest arc-welding processes. It is very versatile and is widely used in industry. A major advantage of stick welding is that it generates its own shielding to protect the weld deposit from atmospheric contamination. Most welding codes allow the process to be used in winds up to about 20 mph. This is one reason it is widely used when welding outdoors.

MIG welding is more efficient and is used more often on an annual basis. However, MIG cannot tolerate wind beyond about 5 mph without the use of an external windbreak. Most welding codes limit the use of gas-shielding processes, such as MIG welding, to winds of 5 mph unless a windbreak is used.

Another reason for the popularity of stick welding is its simplicity. In addition to the stick electrode itself, you only need a power source and a simple electrode holder. The welding power cables can be very long, extending from a power source that might be located in a truck or a hundred or more feet from the work location.

Stick welding might also be considered for its access to the weld joint. The long electrode makes it a usable welding process when no other method is accessible. The coated electrode is insulated on the sides and can be inserted next to a metallic part and a weld placed 5 or 6 inches from the stick electrode holder. In some pipe welding joints, where access is very limited, weldors may even bend the electrode to make a short section of weld. They use only the end where the flux did not peel away.

Fig. 5.1. Stick welding is one of the oldest arc welding processes. It was developed in the early 1900s. It is very versatile and is widely used in industry. A major advantage of stick welding is that it generates its own shield to protect the weld deposit from atmospheric contamination. It can be used when wind up to about 20 mph is present.

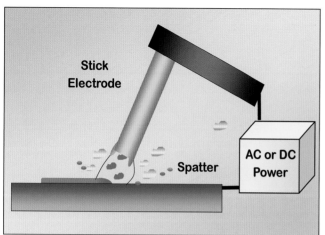

Fig. 5.2. The stick electrode is the heart of the process. It consists of a center-core rod coated with a flux. The flux is a mixture of fine-powdered ingredients mixed with liquid silicates to provide a binder. The flux is extruded around the center-core rod and baked in ovens.

Electrodes

The heart of the stick-welding process is the electrode itself. It consists of a center-core rod coated with a suitable flux. If welding steel, this is a simple steel rod ranging from 3/32 to 1/4 inch in diameter. A 3/32-inch diameter is a common size for welding thin materials.

Fig. 5.3. A stick welding power source can be very simple, such as this old standby, which is still sold. This particular power-source design is available as an AC or combined AC-DC version. They typically weigh from 150 to 175 pounds.

The flux is a mechanical mixture of fine-powered ingredients that are mixed with liquid silicates to provide a binder. This mixture is extruded around the center rod. This coating usually contains several ingredients: a.) alloys to improve weld strength and toughness; b.) deoxidizers, such as ferromanganese and ferrosilicon, that prevent excess oxygen from creating carbon monoxide gas porosity; c.) slag formers, such as rutile and silica, that protect the hot solidified weld by producing a slag, which also aids in forming smooth, well-wet beads; d.) carbonates to produce carbon dioxide gas for shielding the molten weld metal; and e.) arc stabilizers, such as the liquid potassium silicate binder, to help start and maintain a stable, steady arc.

After extrusion around the center rod, the liquid contained in the binder is removed by baking in ovens. The coating is removed for a short distance on one end to allow it to be clamped in an electrode holder so welding current can be conducted through the center core rod.

As stick-welding progresses, the outer coating melts and provides the shielding atmosphere and the deoxidizers to chemically combine any oxygen present into harmless oxides.

The slag that forms covers the molten weld as it solidifies. The molten slag/molten weld metal interface helps provide a smooth well-shaped weld bead. Some of the oxides float to the top of the weld deposit with the slag. The slag stays molten as the weld solidifies and continues to protect the hot weld surface from oxidation.

Numerous types of electrodes are available to weld mild, alloy steel, stainless steel, and hard-facing materials. Materials are also available for welding cast iron and nickel alloys, and some are available for welding aluminum, although this is not a preferred method.

Equipment

Conventional transformer designs are powered by 220-volt alternating-current power lines. These can provide AC or DC/AC welding power. Larger industrial stick welders are available with up to 600-amp capacity. Inerter-based designs are much lighter and can be easily moved from job to job.

Conventional Power

Stick-welding power sources can be very simple, such as the old standby seen in Figure 5.3, often referred to as the tombstone because of its shape. This particular power source can be purchased as a simple AC welder. AC welders are sometimes called buzz boxes because of the sound they make from the vibration of the current control mechanism that varies the welding current.

AC welders are also available in combination AC/DC versions. They typically weigh 150 to 175 pounds. The large internal transformer converts the 220-volt line input into welding power that may be 80 volts when not welding. And it has an output that follows a curve down to 0 volts when set at 150 amps, for example. This type of power output is called constant current because the voltage can vary as the arc length changes, but the current remains at about the preset level. Because electrode burn-off rate and weld penetration are predominately dependent on current, a weldor can produce a uniform bead.

In comparison to AC, DC power provides a more stable arc, particularly with low-current, small-diameter electrodes. DC reverse polarity, also called electrode positive, provides the best mechanical properties, specifically weld metal toughness. With reverse polarity, the arc is most stable and can be held at a short distance. A shorter arc length yields less contamination from air and produces a lower level of oxides and nitrides, which detract from toughness.

Arc Blow

This condition can occur when welding steel with DC power. During arc blow, the arc may move to one side or backward and influence weld quality and appearance. If the arc bends back, the weld is humped in the center and undercut on the edges. If it moves rapidly from side to side or randomly, it could pull air into the weld puddle and cause porosity. With DC power, the current is moving in one direction, creating a magnetic field that is circular around the electrode and arc. Any magnetic field in the base material causes the arc to move. The current flows in both directions and usually eliminates the problem of preferential magnetic fields and avoids the problems caused by arc blow.

Fig. 5.4. Even higher amperage stick power, such as this 650 amp unit, is available and also used for carbon arc gouging. In carbon arc gouging an arc is established between a carbon electrode and the workpiece, and a high velocity air blast blows the molten metal away and produces a gouge. It is used to remove defective welds and prepare weld joints.

For industrial applications, stick power sources are available in capacities of 400 amps and higher. Some models have a 400-amp capacity and uses advanced electronic technology to improve arc performance. However, it still contains a large transformer and weighs about 400 pounds. It has a rugged, galvanized, bottom chassis and large rear wheels for easy transport.

Inverters

The major advantage for stick welding is the ability to make small, lightweight products. The 200-amp stick power source shown in Figure 5.6 is an inverter-based design (see

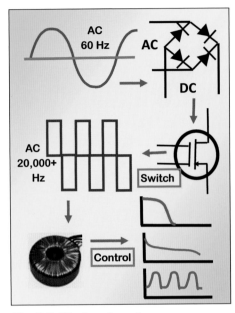

Fig. 5.5. Modern inverter power supplies have many starting and performance features not attainable with conventional older transformer designs. The large conventional transformers are replaced with ones weighing perhaps 20 times less. The purple arrows show the sequence of components employed. The use of microprocessor-based controls allows very accurate monitoring and control of the arc performance.

Chapter 4 for details on the operation and benefits of inverter-based power in relation to TIG welding). It weighs less than 25 pounds, so it's very portable. Arc-starting characteristics are also improved. This particular unit is very efficient, requiring only a 16 amp input fuse. It operates at 99-percent power at full load.

Stick-welding power is also available with generators powered by Diesel or gasoline engines. These are often installed in the back of a truck. The ability to use long cables allows the stick process to be used on construction sites, etc. These engine-powered welders are available from 200- to more than 500-amps capacity.

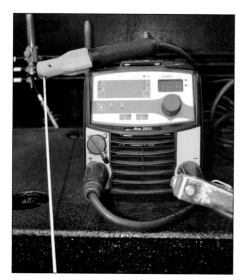

Fig. 5.6. This 200-amp stick power source is an inverter-based design. It weighs less than 25 pounds, making it very portable. Arc starting characteristics are also improved. This particular unit is very efficient, requiring only a 16-amp input fuse. It operates at 99-percent power factor at full load.

Fig. 5.7. Stick welding power is also available with generators powered by Diesel or gasoline engines. These are often installed in the back of a truck. These engine-powered welders are available from 200 amp to more than 500-amp capacity. This Miller Trailblazer 302 Air Pak welder includes 13,000 watts of auxiliary power and can deliver 26 cfm of compressed air.

Purchasing Stick Power

Unlike with oxyacetylene equipment (where the gas source and cylinder refilling options must be addressed with the gas supplier), stick welding uses no gas. The equipment can be obtained from a number of sources, including DIY stores such as Lowes and Home Depot.

When considering a product from discount tool stores, such as Harbor Freight, you need to check the duty-cycle rating. Some may be rated at only 20 percent, which means that for every 10 minutes of use it can only be operated for 2 minutes. If operated longer a thermal switch may shut off the power source until it cools, or worse could cause a failure. These low-usage welders may be fine for small home projects but not for welding heavier plate or when a significant amount of weld is required.

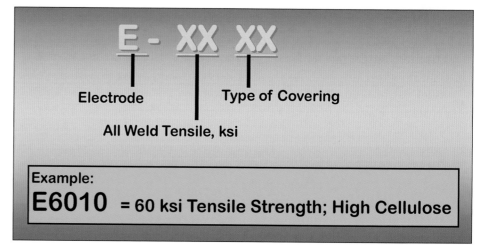

Fig. 5.8. AWS has a designation system for the many types of stick electrodes available. Standards define electrodes for steel, stainless steel, nickel, and aluminum. Part of the designation for steel electrodes, the first two-digit number is the minimum weld tensile strength, in ksi.

Industrial welders have duty-cycle ratings of 60 percent or higher. When stick welding, a 60-percent duty cycle is generally sufficient because stopping occurs as you make the welds, cleaning slag from previous weld beads before proceeding. Changing to a new electrode generally consumes 40 percent of the time involved in making welds.

Filler Metals

The American Welding Society has a designation system for the many types of stick electrodes available. Standards define electrodes for steel, stainless steel, nickel, and aluminum. However, stick welding aluminum has a number of disadvantages and is not widely employed.

All Listed Require Minimum Impacts of 20 ft-lbs @ -20 F E7024 Requires 17% Elongation; Others 22% Elongation		
AWS A5.1 Classification	**Covering**	**Type of Current (Positions Allowed)**
E6010	High Cellulose with Sodium	DCRP (F, V, OH, H-fillet)
E6013	High Titania with Potassium	AC , DCRP, DCSP (F, V, OH, H-fillet)
E7018	Low Hydrogen with Potassium	AC or DCRP (F, V, OH, H-fillet)
E7024	Iron Powder with Titania	AC, DCRP or DCSP (F, H-fillet)
F= Flat Postion; V = Vertical; OH = Overhead H= Horizontal Fillet		

Fig. 5.9. AWS designations for four common electrodes: the first two digits define the minimum weld tensile strength at either 60 or 70 ksi. The last two digits, for example the 10, must be referenced in a standard chart and indicate the type of coating and the power that must be used. In this case, 10 defines that it is usable with DC reverse polarity and can weld in flat, vertical, overhead, and for horizontal fillets.

Current Range for Two Classes of Electrodes vs Size		
Electrode Size	**E6010 Amps Range (Optimum)**	**E7018 Amps Range (Optimum)**
3/32	40-75 (75)	70-100 (90)
1/8	80-140 (100)	890-160 (120)
5/32	130-175 (140)	130-220 (120)
3/16	150-210 (160)	200-300 (120)
7/32	180-275 (190)	250-350 (120)

Fig. 5.10. There is an optimum current range for each type and size electrode. The two types that have significantly different coatings are shown here. It is important not to exceed the ranges listed because the current heats the electrode and excess current creates too high a temperature.

The designation for steel electrodes is the same for most types. The two-digit number following the E (for electrode) is the resulting minimum tensile strength, in ksi, of a weld made with the designated electrode. This strength is determined from a sample machined from an all-weld metal, multipass weld deposit that does not have significant dilution into the base material. The last two digits are very important but a specification table must be consulted to define the meaning.

Figure 5.9 shows four common steel welding electrodes. For example, E6010 tells you the tensile strength of a weld made with the electrode is a minimum of 60 ksi and that it is usable with DC reverse polarity and can weld in flat, vertical, overhead, and horizontal fillets. Although stick welding is not the ideal process for welding sheet metal, when needed, a 3/32-inch-diameter E6013 can operate as low as 40 to 50 amps.

An E6013 is the best choice if stick welding is used at low currents.

The bottom electrode in Figure 5.9 is a 7024. The 70 specifies an all-weld-metal tensile strength that exceeds 70 ksi. The coating includes iron powder and titania (often referred to by the titania-containing mineral used, called rutile). Note that it can be used with AC as well as DC power with either reverse or straight polarity. However, it is limited to welding in the flat position or for making horizontal fillets. These AWS designation numbers are marked on the electrode coating.

There is an optimum current range for each type and size of electrode. The two types that have significantly different coatings are shown in Figure 5.10. It is important not to exceed the range listed because the current heats the electrode and excess current creates too high a temperature.

Mechanical Properties for Two Classes of Electrodes		
	E6011	E7018
Tensile	76 ksi	78 ksi
Yield	66 ksi	68 ksi
% Red. Area	56%	75%
% Elong.	22%	30%
Charpy Impact	31 ft-lbs @ -20 F	168 ft-lbs @ -20 F

Fig. 5.11. Mechanical properties are excellent for the low-hydrogen E7018 class of electrodes. Impact properties of 168 ft-lbs ensure resistance to brittle failure. Other strength levels and different coatings with low-hydrogen classifications are available.

Mechanical properties are excellent for the low-hydrogen E7018 class of electrodes. Impact properties of 168 ft-lbs ensure resistance to brittle failure. Other strength levels and different coatings with low hydrogen classifications are available.

Typically, unopened containers of electrodes can be stored safely for years under normal dry-storage conditions. They should be protected from humid air after the container is opened by being kept in a sealed metal or plastic container or in a rod oven set at 225 to 300 degrees F. If needed, they can be reconditioned by baking for one hour at 700 degrees F. When moisture is absorbed into the electrodes, it is chemically combined so simply heating to 250 degrees F, for example, does not lower it to acceptable levels, and therefore 700 degrees F is needed.

Learning Stick Welding

Stick welding is not commonly used in automotive work, and you may not have the skills needed to use it properly. Therefore, following is some basic information and exercises that can be used to make stick welds.

If you have a TIG welder, all you need to stick weld is an inexpensive electrode holder. It is worth the effort to develop some stick-welding skills because there may be good reasons to use the process. For example, repairing a crack in a cast-iron part can be done with a special electrode designed for that purpose. Also, if a special stainless, such as a type 316, needs to be welded, purchasing a small package of stick rods may be a far less expensive option than purchasing special spools of 316 stainless MIG wire or a box of TIG rods.

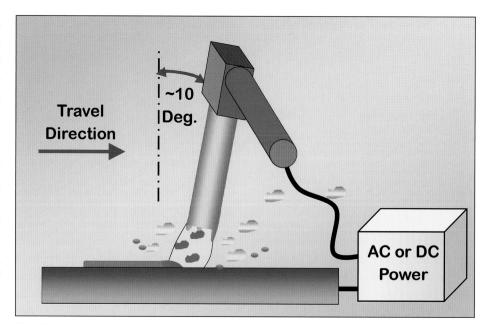

Fig. 5.12. Stick welding is not commonly used in automotive fabrication but may find uses that uniquely match the process features. Some learning exercises are provided to enable making stick welds, if needed for example, to repair cracks in cast-iron parts, or for emergency repairs at the racetrack.

If you already have TIG welding skills, stick welding is easier to learn. Some of the differences should be practiced until you are comfortable making quality welds.

Basic Practice

Assuming you have a stick-welding power supply or a TIG welder and an electrode holder, start with some 3/16-inch steel plate and 3/32-inch-diameter E6013 electrodes. An E6013 electrode has a high-titania (rutile) coating and uses potassium, which improves arc stability.

An E6013 provides a soft, steady arc with low spatter and is easy to restrike. Although an E6013 electrode can be used with AC or DC power and either straight or reverse polarity, it is recommended to start with DCRP, also referred to as electrode positive.

Starts: Making "starts" is different from starting a TIG weld where high frequency is used. With stick welding, it is necessary to strike the end of the 14-inch-long electrode as if you are striking a match. Begin with the optimum current for a 3/32-inch-diameter E6013 of about 80 amps. Following the power source manufacturer's instructions on setting the power source.

Next, try striking and immediately breaking the arc by lifting the electrode tip from the plate. Do not be surprised if the end of the electrode welds itself to the plate. This is probably the most frustrating part of learning to stick weld. Don't give up; you'll get the hang of it with practice. The long, 14-inch length makes this more difficult than scratch starting TIG. Repeat this exercise at least 25 times before attempting to make a weld.

Remember, restricting is more difficult since the end of the electrode has some slag over the melted end. This must be broken with a sharp, hard, quick movement. As long as the tip is moving when it strikes the plate, it does not stick. The trick is to have it move and stop as soon as an arc is established so the tip is not pulled away and the arc extinguished.

Short Welds: When capable of starting and restating the arc, begin holding the arc for a longer time. When the puddle forms, move the electrode slowly along the plate (at a rate of about 2 to 3 inches per minute). During the movement, try to keep the arc length at about 1/8 inch long. As the electrode burns back it must be moved as it progresses along the joint. That requires practice to achieve a smooth, uniform weld deposit. This will become automatic with sufficient practice.

Try making a weld across the plate. Note that for all except vertical welding, use a 5- to 15-degree drag or backhand angle (see Figure 5.12). This helps shield the weld puddle. If vertical welding, a forehand angle is needed to prevent the metal from dropping.

Sound is another way to monitor the arc length. The correct arc length gives a steady cracking sound, like frying eggs. If the arc becomes too long, the cracking reduces and more of a humming sound develops. This sound is difficult to maintain as the arc is trying to extinguish. If the arc becomes too short, popping sounds are heard and the electrode tries to stick to the plate.

The arc length, when stick welding, is harder to see than a TIG arc length. A 1/8-inch-long arc should be maintained while the electrode is melting back. The tip of the electrode must move while it travels along the joint.

Try a somewhat longer arc length and observe the bead shape.

Hand Control: My old friend Butch Sosnin suggests an interesting way to practice hand/eye coordination without having to weld. It is not a substitute for welding but can be done anywhere to help develop and practice the movement skills needed

for stick welding as well as TIG welding (see Figure 5.13).

You use a heavy, 1/8-inch-thick washer and a sharpened pencil. Place the washer on a piece of white paper. Holding the end of the pencil, move the washer along a straight line, but a.) do not allow the pencil to touch the paper and b.) don't allow it to move above the top of the washer and lose the movement. When the 1/8-inch washer is mastered, try reducing to a 3/32-inch-thick washer.

Bead-on-Plate Welds

Following are the steps toward proper procedure.

1. Start at the end of a plate and make a full-length weld. Watch the arc and try to keep its length constant. As the electrode moves across the plate, observe the molten puddle. Try to watch the whole puddle and observe the action of the molten metal. Pay attention to the leading edge as it melts the base material. Try using the whole electrode until the stub is down to about 2 inches. Finish the weld with a new electrode as needed.
2. Clean this first weld with a chipping hammer and wire brush. Look at the edges to see where you deviated from a straight line.
3. Next, place another bead about 1/8 inch from the first. Continue until that weld bead is finished.
4. Adjust the power supply to reduce the current about 10 amps. Make a weld so you can see or hear a difference in the arc. After making a full-length bead, reduce the current another 5 amps. The difference should now be obvious. After another bead and reduce the current another 5 amps.

Fig. 5.13. My old friend and weldor trainer suggested an interesting way to practice hand-eye coordination without having to weld. It is not a substitute for welding but can be done anywhere to help develop and practice the movement skills needed for not only stick but also TIG welding. It simply involves moving a washer across a white sheet of paper with a pencil, without marking the paper or losing the washer movement.

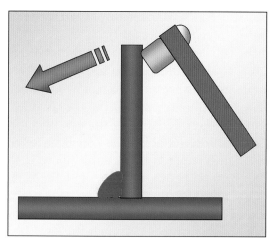

Fig. 5.14. Now for the real test of your first stick fillet weld: break it open. Place the horizontal plate in a vice, strike the vertical plate, and bend the weld away from the unwelded side with a heavy hammer. It may take several blows, but try to flatten the vertical member on to the horizontal member. Once the weld is broken, examine the weld surface for porosity, lack of fusing to the joint apex, or other defects.

Whipping the Bead: You can try this technique when it becomes necessary to handle gaps in joints and when welding thinner material. Because welding speeds are relatively slow, as the weld progresses the temperature in front of the weld increases as well as where the weld was solidifying in the rear. This causes the metal to melt faster and to increase the size of the puddle. Whipping is a technique to help control the heating rate of the weld puddle and control puddle size.

1. After the arc is struck, hold the electrode still for a second.
2. Then move the electrode about 1/4-inch forward and increase the arc length slightly.
3. As the rear of the puddle is about to freeze, move the electrode halfway back from the leading edge to the center of the puddle.
4. Repeat the sequence.

This whipping motion should be done with the wrist, not by moving the arm.

The finished weld bead shows the stepped solidification pattern.

Weaving the Bead: Now try another technique useful for bridging gaps and controlling the weld puddle heat.

1. Move the arc from one side of the intended weld joint path by about 3/32 inch. Hold the arc there for about a second.
2. Then move the arc to the other side about 3/32 inch while moving it forward about 3/32 inch.
3. Repeat this while progressing along the intended weld path.

Make a series of weld beads using weaving and the whipping technique until the welds are fairly straight and uniform.

Horizontal Fillet Weld

It is time to make a weld between two plates. One type is a horizontal fillet weld.

1. Tack two 3/16-inch plates at 90 degrees and make a fillet joint.
2. Place the electrode at a 45-degree angle directly in the apex of the joint. As with your practice beads, a slight backhand or drag angle of about 10 degrees should be used.
3. Move the electrode along the joint at a steady pace. The speed should be adjusted so that the weld wets the vertical leg of the fillet. The slag helps hold the weld bead on the vertical member.
4. After the weld has been made, remove the slag and wire brush the surface. Look for uniform edges and undercut.

Now for the real test: Break the weld open. Place the horizontal plate in a vice and with a heavy hammer, strike the vertical plate and bend the weld away from the unwelded side (see Figure 5.14.) It may take several blows to flatten the vertical member onto the horizontal member. It may require reversing the bent plate and bending it several times to get it to break. Once the weld is broken, examine the weld surface for porosity, lack of fusing to the joint apex etc.

S·A DESIGN

MIG WELDING

MIG welding is the leading arc joining process, and more than half the weld metal deposited in the United States is MIG wire. MIG welds can be found on components such as car and truck frames, axles, driveshafts, mufflers, catalytic converters, suspension components, and wheels. Skilled weldors using modern MIG equipment help produce heavy earth-moving equipment, agricultural tractors, and submarine hull joints. With the use of the appropriate welding wire, MIG-weld deposit strength, ductility, and impact properties can exceed those of high-strength, high-toughness steels. The process is widely used for welding steel, aluminum, and stainless steels.

There are several variations of MIG welding and the basics of how the process works is must-have information. You also need to understand why MIG wire melts and what parameters affect weld penetration. These are critical factors that control weld quality.

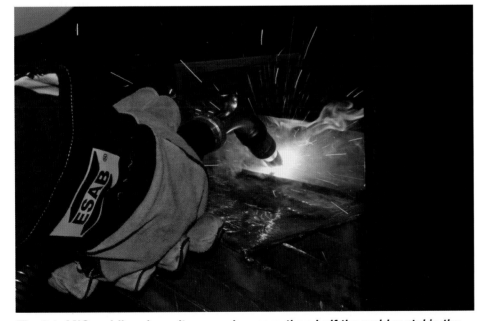

Fig. 6.1. MIG welding deposits comprise more than half the weld metal in the United States. The process can produce welds exceeding the strength and toughness of high-strength steels used for critical applications such as submarine hulls.

MIG WELDING
Process Variations:
> Spray Arc

> Short Arc

> Pulsed Arc

Fig. 6.2. There are three significant variations of MIG welding. Each has a place in automotive welding. The most commonly used variation for automotive-type structures is MIG short arc welding. It can weld at low currents for bridging joint gaps and for welding sheet metal.

Process Variations

Spray arc, short arc, and pulse arc are three major MIG welding processes, and each has its application in automotive fabrication. Most MIG welding used for street rods and race cars employs solid wire. An alternate to solid wire is cored wire.

Spray Arc

Spray metal transfer, or spray arc, was the first MIG process, introduced in the 1950s. In this mode, the metal drops leave the tip of the welding wire in a very fine stream and cross the arc. About 98 percent of the material becomes part of the weld deposit. There is little spatter and that which occurs is very fine and can often be wire brushed from the surface of the base material.

Spray metal transfer can only exist with shielding gas that has a high percentage of argon. Two shielding gases used for welding steel—2-percent oxygen and 98-percent argon or 5-percent oxygen and

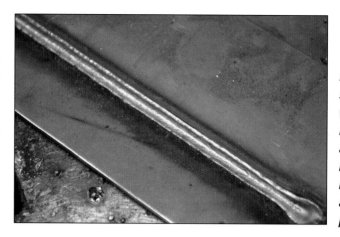

Fig. 6.3. A smooth surface with uniform well-wet-in edges identifies the spray arc weld. There is very little spatter; the surrounding plate is clean and often requires no post-weld cleaning

95-percent argon—produce stable spray-arc metal transfer.

For a given wire diameter, a minimum current is needed to produce spray droplet transfer. For example, spray transfer with .030-inch-diameter wire occurs above 150 amps. The voltage needed is also relatively high, and the combination produces a very hot arc. It is too hot to be used to join thin sheet metal. With .035-inch-diameter wire the minimum current to achieve spray arc is 165 amps with a 98 percent argon and 2 percent oxygen shielding gas.

Before techniques were developed to MIG weld at lower currents, wires as small as .020-inch in diameter were employed for some applications. As you can imagine, feeding these small-diameter wires was very difficult.

Short Arc

If spray-arc transfer were the only MIG welding process, MIG would not have dominated the industry. However, it was discovered that another method of metal transfer could be used. It was called short-circuiting metal transfer, or short arc, and it has gained wide acceptance.

The wire stubs into the weld puddle, and the arc extinguishes when the arc voltage is lowered to a level below that needed to maintain spray arc conditions. If the welding power source used a simple car-battery-type output, the direct short would cause the amperage to dramatically increase when the wire touched the molten weld puddle. As a result, the length of wire from the contact tip to the weld puddle would melt explosively.

To prevent this rapid rise in amperage, a large inductor can be added in the welding circuit of conventional welding machines. The inductor limits the current increase rate, allowing

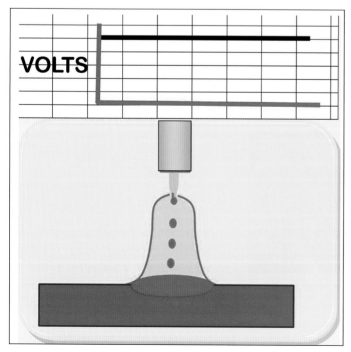

VOLTS

Fig. 6.4. When the MIG process was invented, spray arc was the common metal transfer mode. This graph shows the very stable, uniform situation MIG provides. Fine metal drops are pinched from the end of the wire in a fine stream. There is very little spatter and 98 percent of the metal reaches the weld puddle.

Arc Length

I recall the first time I welded with MIG and wondered how the arc automatically remained at a fixed length. I had just practiced welding with stick electrodes for which arc length was maintained manually, and I thought the wire feed speed might be controlled and varied to maintain the proper arc length.

But the matter remained a question until the following summer when I worked at a welding research and development laboratory. At that time, I learned that the self-correcting characteristics of the power supply automatically maintained MIG arc length. In Figure 6.4 the simplified graph of voltage, with time displayed above the schematic of the spray arc process, shows what is referred to as a constant potential output of the MIG power supply.

Unlike stick welding that uses a constant-current power supply output, the voltage of MIG power systems remains essentially constant regardless of the welding current. It is similar to the output of a car battery but with adjustable voltage. To maintain the arc voltage, amperage rapidly increases if the arc length attempts to shorten causing more wire to melt and arc length to increase.

Conversely, if the arc gets longer than the available voltage can sustain, the amperage quickly decreases, melting less metal and the arc length shortens. This happens very quickly and is not visible to the naked eye. There is no electrical circuit adjusting the wire feed rate to maintain a constant arc length. The wire just feeds at a content preset rate and the constant-potential power source automatically provides the self-correction of arc length. ■

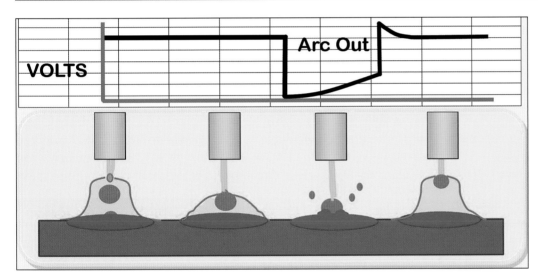

Fig. 6.5. Short arc allows the MIG process to be welded out of position and is useful for welding sheet metal. The arc is extinguished at a high frequency as the wire touches the molten metal. The arc voltage and power-supply characteristics control the amount of time the arc is on or off.

time for the molten drop on the wire tip to be pinched off and pulled into the weld puddle by surface tension. Therefore, a relatively non-violent metal transfer occurs.

When the molten drop detaches, the arc is re-established with the assistance of additional energy stored in the inductor during the short. A small amount of metal is ejected in the form of spatter.

With MIG short arc, the voltage and current are not limited to the minimums needed for spray arc. The

majority of automotive MIG welding covered in this book is done in the short-arc metal transfer mode.

To achieve stable short-arc welding conditions, the shielding gas must contain a minimum of 8-percent carbon dioxide, but a mixture of 22- to 25-percent carbon dioxide in argon is commonly used. The use of 100-percent carbon dioxide can provide short-arc metal transfer, but with variable size, larger molten drops, and more spatter. Some of the spatter sticks to the surface of

the welded plates and often requires removal after welding.

With the proper power source and shielding gas more than a hundred short circuits can occur per second. Don't expect to see the short taking place. The size of inductor and other variables determine the percentage of time the arc is on or off.

Pulse Arc

MIG pulse arc is a more complex process than short arc for lowering

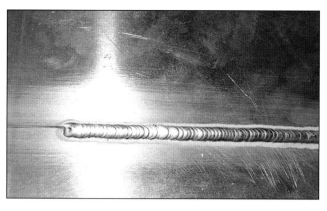

Fig. 6.6. Short-arc welds are often identified as having a visible solidification pattern. The arc is typically manipulated in a systematic manner to ensure it is melting the base material and achieving good penetration.

total power input compared to standard spray-arc welding. In fact, it's a spray metal process that requires lower heat input. First developed in the early days of MIG welding, a microprocessor is required to control the fast-switching power sources to make this mode of metal transfer a more practical welding tool. Modern pulse-arc MIG welders operate at frequencies in excess of 20,000 cycles per second, but do not rely on short-circuiting to lower the average current and voltage needed to produce stable, fine-droplet metal transfer. The average welding voltage and cur-

rent are set well below the needed spray-arc minimums.

Before the metal droplet at the end of the wire has time to grow and short to the molten weld puddle, the voltage and amperage rapidly rise for a very short time. The rapid spike in welding power causes a molten metal drop to be pinched from the wire tip. The molten drop transfers in a spray mode, but at a much lower average voltage and current than occurs when spray-arc welding with a simple constant-potential power source.

Pulse-arc MIG welding allows the use of relatively low average cur-

rent and voltage with larger wire diameters. For example, when welding steel, the average pulse-arc current can be as low at 50 amps with .035-inch-diameter wire. This compares to the 150-amp minimum with a conventional power source to achieve spray-arc transfer.

When welding aluminum, the minimum spray-arc current is 135 amps for .045-inch-diameter wire. That can be too hot for thinner materials or out-of-position welding. However, the minimum average current for pulse arc can be as low as 45 amps. Even 1/16-inch-diameter aluminum wire can weld with an average pulsed-MIG current of 85 amps. It is far easier to feed large-diameter wires in a normal MIG gun system than smaller-diameter wires, especially aluminum ones.

Unlike MIG short arc, pulse arc can be performed with almost no spatter, and the weld deposit can look similar to TIG welds. However, to make welds of this quality, you need to set the proper pulse peak

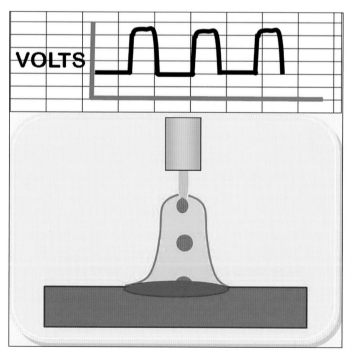

Fig. 6.7. Pulse-arc welding operates at a similar low heat input as does short arc, but without short circuits. Rapidly raising the voltage and current in millisecond-length pulses pinches molten drops off the wire. The drops are small and uniform with little resulting spatter.

Fig. 6.8. A properly tuned pulse-arc MIG system made this fillet weld. The weld appears similar to a spray arc weld, having a well-wet-in and smooth surface with good penetration. Made with low average heat input, pulse arc is ideal for welding out of position.

voltage and current, pulse frequency, base or background voltage, and have very fast current-rise-time capability in the power source. These variables are dependent on wire type and gas mixture. The use of microprocessor-based control allows selecting the proper welding conditions from those stored in memory.

Cored Wires

Small-diameter solid wires (steel, aluminum, and stainless steel) are used for the majority of MIG welding in street rod and race car applications. Flux-cored and metal-cored wires are other types of wires often used in industrial applications on mostly thicker materials. They are welded with similar shielding gases as utilized with solid wires.

Also available are self-shielding flux-cored wires that operate without shielding gas. Self-shielding wires contain ingredients in the center core to counter the potential problems with nitrogen and oxygen in the air and provide some shielding similar to stick welding. These are often found packaged with lower-cost MIG welders. Eliminating the need for shielding gas makes them, on the surface, appear attractive. However, the small-diameter, self-shielding wires needed for welding thin material create a good deal of smoke and a large amount of spatter and slag.

For welding sheet metal, these self-shielding cored wires do not produce the best results, and the use of these products' false economy may create a bad impression of what welding can accomplish, so I recommend avoiding them. However, if welding other than sheet metal such as 3/16-inch and heavier steel, these products

Cored Wire

1) Flux Cored Wire
2) Metal Cored Wire
3) Self Shielded Flux Cored Wire

Fig. 6.9. Cored wire is used for some MIG welding and is usually made from a strip of metal formed into a U shape and then filled with iron powder, arc stabilizers, and flux ingredients. The filled U is then formed into a tube and drawn into the wire. Most flux-core and metal-core wires are welded with similar shielding gases as with solid wires. Some wires identified as self-shielding include ingredients that form some gaseous products to help protect the weld from air contamination.

may be successfully applied, particularly with the price of shielding gas increasing significantly in recent years. If you wish to aspire to a "pro" status, however, MIG welding with shielding gas is often preferred. When considering equipment to purchase, it is advisable to obtain a welder that has the ability to use gas; even if you start welding with self-shielded wire, you can upgrade to gas shielding.

One other consideration is the size of the self-shielded wire. As with solid-wire MIG, where occasionally .023-inch diameter is used, the smallest size wire I prefer is .030. It makes feeding easier and can weld at higher currents to get good penetration on materials thicker than sheet metal. Self-shielding wires for small welders are generally available in .030 and .035 sizes. Feeding these tubular wires is generally more difficult than solid

wire. I prefer the .035 size, which should be useful for all but thin sheet metal and as noted, weld quality on thin material is superior with solid wire and gas shielding and generally only should not be considered optimum with self-shielded wire.

In summary, for industrial welding on heavier sections, gas-shielded flux-cored wires and metal-cored wires have some unique advantages. Metal-core wires, for example, can contain elements such as sodium or potassium in addition to iron powder in the center core that makes the arc more stable. They can also improve weld bead edge wetting, making it easier to produce quality welds. They are also more forgiving to wire-manipulation errors, etc.

Currently, these products are made in larger sizes than are useful for sheet-metal applications. Some self-shielding flux-cored wires are available that provide good performance for welding thicker steel structures, such as bridges. In addition, these wires can be used outdoors in moderately windy conditions. They are generally available in sizes of .045 inch and larger diameters; they are not made in the smaller sizes for general use on thinner material. They also require careful control of wire feed and voltage to provide the strength, ductility, and durability for welding critical components, and as mentioned, this self-shielding flux-cored wire is generally not suitable for welding thinner automotive components.

Gun Position

Accurately describing how to position the MIG gun relative to the direction of travel can be somewhat confusing. In addition, weldors have

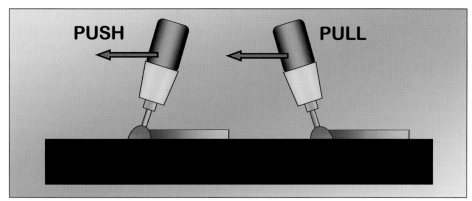

Fig. 6.10. There are a number of ways to describe the angle of MIG gun travel. Push and pull are probably most descriptive. The AWS definitions are backhand for pull and forehand for push. The use of a pull angle is preferred although up to 15 degrees in either direction produces about the same weld result. At times, access to the joint requires a push angle.

differing and sometimes conflicting opinions on how best to hold the MIG gun. Quoting the *AWS Handbook*: "The electrode angle normally used for all positions (of MIG welding) is a drag angle in the range of 5 to 15 degrees." (But not all skilled weldors agree!)

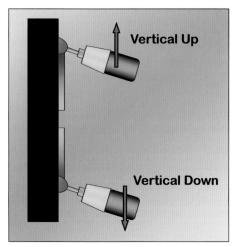

Fig. 6.11. When making a vertical weld, the preferred direction is to move upward. If it's necessary to weld downward, be very careful to ensure the arc is on the leading edge of the weld puddle. If attempting to short arc a vertical weld downward, the weld puddle can easily roll ahead of the arc and prevent penetration into the base material.

Graphic representations of MIG gun angles are presented in Figures 6.10, 6.11, and 6.12. Generally, for downhand welds, a push or pull angle of 15 degrees in the direction of travel does not make a great deal of difference. At times, the angle used depends on the part and access.

Officially, the word backhand is the preferred term for a drag or pull angle. The opposite formal term is forehand, but push angle is often used. Frankly, I've never liked the backhand and forehand description because it can be confusing. Therefore, push and pull are used here. Understanding each of the terms

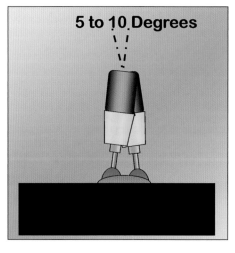

Fig. 6.12. It may be desirable, especially when short-arc welding, to move the MIG gun slightly from side to side across the weld joint. This movement makes a wider weld and helps bridge small gaps. Depending on gun position relative to the joint, a slight twist of the gun handle can provide sufficient movement.

is useful because weldors use all of them, often interchangeably.

When forced to weld in the vertical position it is preferable to weld upward. This ensures the arc can be placed at the leading edge of the weld puddle so melting of the base material occurs. Welding downward is risky since the weld surface may look good but the molten metal may have rolled ahead of the arc and not properly penetrated into the base material. There are times when it may not be possible to avoid welding downward; just be sure the arc is on the leading edge of the molten puddle.

Gun Manipulation

Unlike with some stick welding electrodes where manipulation of the rod tip is preferred, with MIG welding all that is required for many welds is simply moving the gun steadily along the desired weld joint with the shielding gas nozzle held at a uniform height from the workpiece. There are times, however, when a small amount of manipulation may be desirable such as when bridging small gaps. One way to accomplish that is to simply move the MIG gun handle so the welding wire moves slightly across the weld joint. Only a small angular movement is needed.

For some types of weld joints, manipulating the wire tip is desirable. Some weldors also like the resulting weld appearance that manipulation

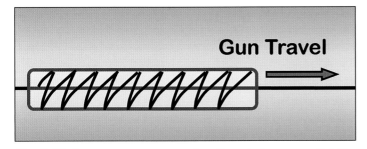

Fig. 6.13. A common movement pattern is to move slightly from side to side while moving forward along the weld joint. When short-arc welding, be sure to keep the arc on the leading edge of the weld puddle to ensure adequate penetration.

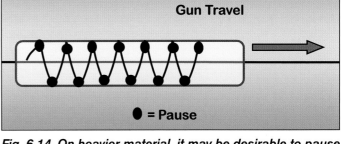

Fig. 6.14. On heavier material, it may be desirable to pause or dwell for a moment on each side of the weld joint. This pause allows the arc to melt the base material at the weld edges and improve wetting. Use a pattern of steady movement forward while pausing on either side of the joint.

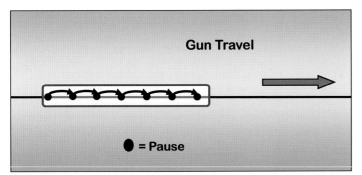

Fig. 6.15. Some weldors prefer to move forward in a straight line and pause slightly before going forward again. This provides a weld appearance similar to TIG welding. This technique creates a distinct weld appearance, avoiding the non-uniformity in weld beads made with straight-line movement that is irregular.

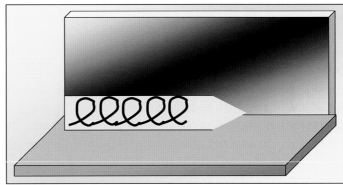

Fig. 6.16. When making fillet welds, the molten metal naturally rolls down due to gravity. Focusing more heat on the upper portion can avoid excess sag or a rolled lower edge. A circular movement pattern can achieve that extra heat concentrated on the upper portion of the joint.

can produce rather than a smooth weld surface. Just be sure the arc is maintained on the leading edge of the puddle. Weaving the gun slightly across the weld can bridge small gaps. In some instances, pausing on one or both sides may be useful.

If making a weld between two different thicknesses of materials, weaving slightly and pausing on the thicker material can produce a better weld. To avoid lack of penetration, the pause time should not allow the weld puddle to flow under the arc.

Moving forward in a straight line and pausing briefly before going forward again is another travel technique some weldors use. When

short-arc welding it's more common to use this step-and-pause technique. This provides a visual weld appearance more typical of TIG welding where it is common to pause when adding filler rod. This technique creates a distinct but uniform weld appearance, whereas slight variations in movement that occur when using

Fig. 6.17. When welding in the vertical upward position, more heat should be focused on the joint edges to ensure good wetting and a flat shape. Moving the arc to the center of the joint ensures penetration into the throat. Moving back slightly and pausing heats the edges.

straight-line travel are seen as non-uniformity in the weld bead.

In horizontal fillet welding, it may also be desirable to spend more

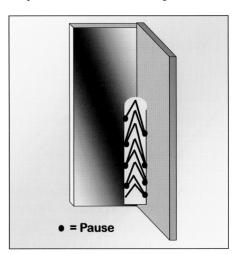

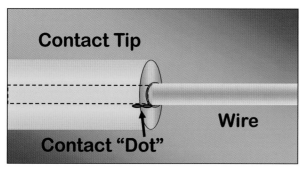

Fig. 6.18. Understanding what causes the wire to melt helps in controlling weld quality. All welding current that transfers from the copper contact tip to the wire occurs at the very end through a small pinhead-sized area. *Although there are several reasons for this current path, the size and conductivity difference between the contact tip and the wire are significant factors.*

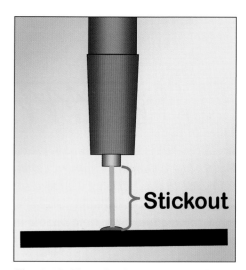

Fig. 6.19. The wire becomes very hot as welding current flows through the small-diameter wire from the end of the contact tip (where it enters the wire) to the arc. It leaves the contact tip essentially at room temperature. As it moves toward the workpiece, resistance heating causes the temperature to rise. By the time it reaches the arc, it might reach 500 degrees F or higher.

time on the upper vertical member to keep the weld metal from sagging and to produce an equal-leg fillet. This can be accomplished with a circular movement. This causes more time to be spent on the upper leg of the fillet, which reduces the tendency for the molten weld puddle to flow downward and cause sagging of the lower leg. The amount of movement needed is dependent on the material thickness and desired fillet weld size.

When welding upward, MIG gun manipulation may be used to produce a quality flat weld. Pausing at the weld edges and using a V pattern assists in ensuring wetting of the weld edges. The metal naturally flows toward the center and drops downward. If the material is thin and only a small fillet size is needed, just a small amount of manipulation, if any, is needed.

What Melts a MIG Wire?

Understanding what causes the MIG wire to melt helps you control weld quality. The first point to understand is how welding current gets into the small welding wire.

Considering the conductivity and cross section area of the copper contact tip versus the steel welding wire, the resistance to current flow is about 350 times more through the

wire and tip. Bottom line? If welding at 150 amps, the current flows to the end of the copper tip before it enters the wire. There are some additional technical reasons, but for practical purposes, all current flows through a small pinhead-size contact dot at the contact tip end.

The small-diameter wire carries all the current from the tip to the arc. The wire would melt explosively if that amount of current lasted for any length of time! Fortunately the wire starts at essentially room temperature and moves quickly, so by the time it gets to the arc, it may reach temperatures in excess of 500 degrees F. It is an efficient means of getting heat into the wire.

A large amount of energy is required to flow electrons into the wire surface and this energy finally melts the tip of the wire. This energy is called the work function. Using the common polarity for MIG welding, the wire connects to the positive terminal of the power supply and the workpiece connects to the negative terminal. The electrons flow from negative to positive and must flow into the already hot welding wire tip. The energy to get the electrons into the wire heats the droplet beyond the melting temperature.

To determine how much metal is melted for a given amperage, see the

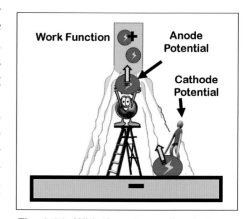

Fig. 6.20. With the wire preheated as it reaches the arc, the energy that causes it to melt is generated at the wire/arc interface. A great deal of energy is required to get electrons into the wire (assuming electrode positive polarity) and this energy is called the work function. The energy expended is what causes the wire to melt and superheat well above the melting temperature.

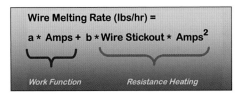

$$\text{Wire Melting Rate (lbs/hr)} = \underbrace{a * \text{Amps}}_{\text{Work Function}} + \underbrace{b * \text{Wire Stickout} * \text{Amps}^2}_{\text{Resistance Heating}}$$

Fig. 6.21. This equation defines the welding wire melting rate. Note that resistance heating is the stickout length (measured as inches) multiplied by the current squared. This has a significant effect on the melting rate.

math equation shown in Figure 6.21. Note the importance of amperage level, which is in both of the equation terms.

The equation has two parts. The first includes a constant (a) that depends on wire type and size, multiplied by the welding amperage.

The second includes another constant (b) multiplied by the wire stickout length (in inches) and amperage squared. This is resistant heating.

Resistance Heating is the energy that is added to the wire when the current passes through it from the contact tip to the arc. The constants for .035-inch-diameter steel wire are .017 for (a) and .00014 for (b). Constants for some other wire types and diameters are presented in the *AWS Handbook* and the original articles entitled "Controlling the Melting Rate and Metal Transfer in Gas-Shielded Metal-Arc Welding" and "The Effect of I²R Heating on Electrode Melting Rate."

For example, if a weld is made at 175 amps with a 3/8-inch stickout, the melting rate is 4.6 pounds per hour.

$$.017 \times 175 + (.00014 \times .375 \times 30{,}625) = 4.6$$

If the stickout is raised to 3/4 inch and the wire feed speed is raised

to maintain 175 amps, the melting rate is 6.2 pounds per hour.

$$.017 \times 175 + (.00014 \times .75 \times 30{,}625) = 6.2$$

Therefore, the longer the stickout, the higher the melting rate. Or the more important consideration, if the melting rate is fixed when the wire feed speed is set, the current is determined by the distance from the MIG contact tip to the workpiece. In our example, if the wire feed speed was kept constant at 4.6 pounds per hour that resulted from the 3/8-inch stickout, and the MIG gun is raised so the stickout increases to 3/4 inch, the current would drop automatically to 143 amps.

Wire Stickout Effect on Penetration

The current level also determines penetration. For this discussion, penetration is defined as distance from the top of a workpiece to the bottom of the weld nugget on a heavy plate. It is one of the most important factors in determining weld quality, especially with short arc welding. Short arc is commonly used in sheet-metal applications and obtaining sufficient penetration is critical to making quality welds.

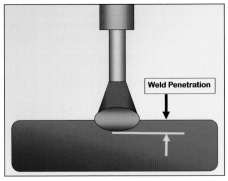

Fig. 6.22. The depth that the weld melts into the base material is called penetration. It is measured by making a cut through the finished bead-on-plate weld and chemically etching the cut surface with an acid that makes the weld structure visible separate from the base material.

Some years ago, Clarence Jackson developed an equation after making and examining hundreds of welds. This work was published in the *AWS Welding Journal*. This equation is useful to gain perspective on the effect of welding variables. The constant defined for .035-inch-diameter steel wire is .0019. Inserting values into the equation for the example .035-inch-diameter wire operating at 175 amps, 23 volts and 10 inches per minute (ipm) travel speed: Penetration = .0019 x [175⁴ amps ÷ (10 ipm x 23² volts)] ·³³³ = .11 inch. Note, the numbers inside the parentheses are manipulated first.

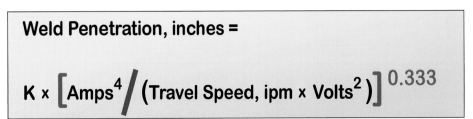

$$\text{Weld Penetration, inches} = K \times \left[\text{Amps}^4 \Big/ \left(\text{Travel Speed, ipm} \times \text{Volts}^2 \right) \right]^{0.333}$$

Fig. 6.23. This equation defines the weld variables that affect the approximate amount of penetration. Of significance is that welding amps are raised to a power of 4 in the numerator. Therefore, when the amperage is raised, it has a very significant affect on penetration. Both travel and voltage are in the denominator, so when these variables increase, the penetration decreases.

Stickout (inches)	Amps	Penetration (inches)	% Loss in Penetration
3/8	200	.127	base
1/2	184	.114	11
5/8	172	.104	18
3/4	162	.096	24
7/8	154	.090	29

Fig. 6.24. This table presents some results of stickout change on melting rate and penetration. It assumes the wire feed speed remained constant. As stickout increases, amperage reduces, which causes weld penetration to decrease. Therefore, to maintain penetration, it is very important to keep the gun tip at a fixed distance from the weld.

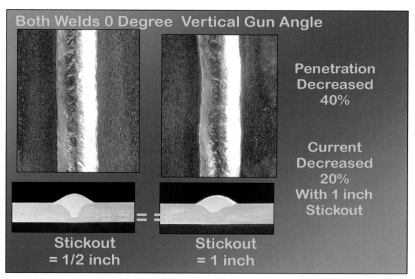

Fig. 6.25. For the welds in this figure, the wire-feed speed, voltage, and travel parameters were all set identically. Stickout was increased from 1/2 to 1 inch, and this caused welding current to decrease 20 percent. As a result, penetration decreased 40 percent.

The value of amps is a very important parameter in controlling penetration. It is raised to the power of 4. Both travel speed and voltage are in the denominator meaning as they increase in value the penetration decreases. However, travel speed is not raised to any power, and therefore it does not have nearly the effect of amps. The increase in voltage (reducing penetration) may not seem obvious. What happens as voltage increases is the arc gets wider at the workpiece, reducing the arc force that creates penetration.

To give you another example of the effect of stickout on penetration, a .035-inch-diameter steel wire sets a base condition using a wire-feed speed to produce 200 amps with a 3/8-inch stickout, and that penetration is .127 inch.

When stickout is increased to 1/2 inch (therefore the MIG gun nozzle was moved 1/8 inch farther from the workpiece), the current decreases to 184 amps, and penetration drops to .114 inch (an 11-percent decrease).

When stickout is moved to 3/4 inch, the current reduces to 162 amps, reducing penetration to .096 inch (a reduction of 24 percent).

Therefore, an increase in stickout significantly decreases penetration, all else being equal. At times, that can be used to advantage. For example if you are starting to burn through or when a joint gap is encountered, pulling the gun back slightly may help.

It is very important to keep the nozzle at a fixed distance from the workpiece, so acceptable penetration and a quality weld is the result.

Examples of Changing Parameters on Weld Results

Figures 6.25 through 6.28 show examples of the effect of changes in welding parameters on weld shape and penetration.

Figure 6.25 shows welds where the voltage and travel speed were held constant. The gun angle was perpendicular to the surface, and the stickout was altered from 1/2 to 1 inch. This stickout increase caused the welding current to reduce 20 percent, which decreased current, and that caused a 40-percent reduction in weld penetration.

Figure 6.26 changes two variables, the pull angle and stickout. The results, as expected with a 15-degree pull angle and a change in stickout from 1/2 to 1 inch, showed penetration decreased 30 percent.

When the pull angle increased to 45 degrees, the overall penetration was low even at 1/2-inch stickout and reduced only 12 percent more with an increase in stickout to 1 inch. A 45-degree angle is very good for making quality welds.

Figure 6.27 shows the effect of travel speed on weld penetration. Remember that the equation defining penetration showed that a change in welding speed should have the least effect. The welds shown here support that finding. Even doubling the welding speed from 10 to 20 ipm only decreased penetration 15 percent.

Figure 6.28 shows the differences. A push angle was used for

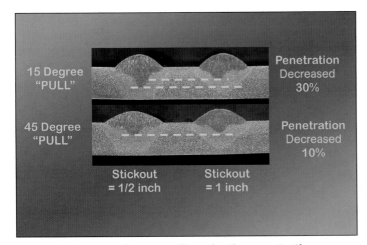

Fig. 6.26. At a 15-degree pull angle, the penetration decreased 30 percent when stickout increased from 1/2 to 1 inch. However, at an excessive pull angle of 45 degrees, even with a 1/2-inch stickout, penetration was low. It only decreased 12 percent with an increase in stickout from 1/2 to 1 inch, but that is from an already low level.

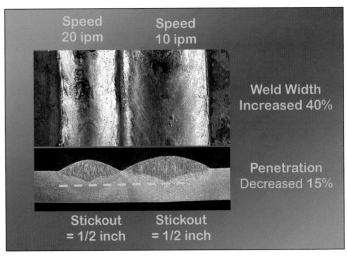

Fig. 6.27. These two welds show the effect of a change in travel speed on penetration. Penetration is affected more dramatically by changes in current, rather than changes in travel speed. In this case, doubling the travel speed only decreased weld penetration by 15 percent.

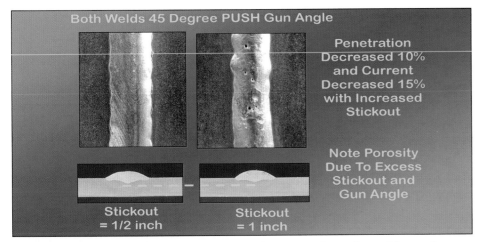

Fig. 6.28. With the gun angle held at a 45-degree push angle, weld penetration decreased only 15 percent with a change in stickout from 1/2 to 1 inch. However, it was at a low level even with the 1/2-inch stickout. In addition, the weld with the longer stickout has visible porosity. The weld gas shield is not adequate with this excess pull angle and large stickout.

both welds. As with a high-push angle weld, there is only a modest reduction in penetration of 10 percent with an increase in stickout to 1 inch. However, note the porosity in the weld bead with the 1-inch stickout. At this poor gun angle, this excessive amount of forward leaning reduced the quality of gas shielding.

Equipment

MIG equipment has advanced in recent years and improved the ability to set and use some of the more advanced processes such as pulsed-arc MIG welding. The following reviews the significant advancements in modern weld power and also discusses

Fig. 6.29. This ESAB Migmaster 215 Pro has a digital volt and amp meter that have a hold feature, so they can be read when welding stops. It also has spot-weld and burn-back timers and a slow inch-down start. A plug is included to allow the easy addition of a spool gun for welding aluminum.

wire feeding, the MIG gun, shielding-gas flow control, and gas waste.

Standard Power Systems

As noted in Chapter 1, the basic equipment needed for MIG welding is a power source, feeding device, and welding gun. The power source can be combined with the feeding device or a separate wire feeder can be utilized.

For many welding applications, the maximum power needed is typically 250 amps. The ESAB Migmaster 215 Pro is a self-contained wire-feed system with a 250-amp capacity. It can hold a wire spool up to 44-pound capacity in its enclosed case. One feature found on the base unit includes the LCD display of amps and volts, which remain on the screen after welding stops for easy reading. It also has an adjustable slow-wire-feed start for improving start quality. In addition, it has an adjustable burnback time, so the wire burns back when welding stops in preparation for the next weld. A spot-weld timer is also built into the system—a very important feature for sheet-metal welding.

Many welds joining sheet metal are best accomplished with a succession of spot welds or short tack welds. Using timed spot welds

Fig. 6.30. In short-arc welding, the inductor's function is to limit the rate that the welding current can increase when the wire shorts to the weld puddle. This allows time for the molten drop to transfer, due to being pinched from the wire and the weld puddle surface tension.

ensures uniform penetration without burning through. The Migmaster 215 Pro also includes a connection for a spool gun, which some prefer to use for aluminum. It can also feed aluminum wire using the normal built-in wire-feed system.

An important component of a quality MIG welder is the inductor necessary to produce quality short arc welds. Looking at the internal construction of the Migmaster 215 Pro, the main transformer and inductor are visible. Figure 6.31 shows the main transformer (on the left), which reduces the 220-volt AC line input to the required welding voltage level the large, copper-wound inductor (on the right side of the metal partition). Equipped with large wheels and a large gas cylinder support shelf

Fig. 6.31. To see what makes a quality arc, you need to look under the hood. On left is the main transformer that converts the 220-volt input AC power into lower welding voltage. On the right is a large inductor. For short-arc welding, this inductor is essential to produce quality arc stability. Low-priced welders may not have a large inductor.

Fig. 6.32. Modern pulse-arc welders have many features, which makes them easier to use and weld quality easier to achieve. The ESAB Aristo welder employs a very fast switching inverter that provides excellent pulse-arc welds. Switching rates exceed 50,000 per second. What does that do for the weld? The arc can be monitored and controlled at a very fast rate to achieve spatter-free spray-arc drop transfer without limitations from the power source.

in the rear, the Migmaster 215 Pro weighs 215 pounds.

A close-up view of the inductor is shown in Figure 6.30. In conventional-design power supplies, it takes a large, heavy device of this type to produce quality short arc characteristics. Before you purchase a low-cost MIG welder, remember that this type of extra device and controls are needed to make quality welds. Some MIG welding machines may have similar advertised power output, but a smaller-than-optimum inductor may be used for cost considerations.

Pulse-Arc Power Systems

Early MIG pulse-arc welders were difficult to set because many variables had to be controlled, making them difficult to use with consistent results. The introduction of microprocessor controls with fast-switching inverters made MIG pulse arc an effective tool.

Today, spatter-free, exceptional-quality welds can be made using a MIG pulse-arc system. It can achieve TIG-appearing welds at high MIG metal-deposition rates.

The ESAB Aristo is an inverter-based MIG pulse-arc welder. It is versatile and can also be used to make short-arc, spray-arc, DC TIG, and stick welds plus perform arc gouging. It is available at three current levels. For most applications, the 300-amp, 100-percent duty cycle unit is more than sufficient.

Selecting the complex pulse parameters (i.e., background current, pulse current, pulse rise and fall times, and pulse frequency) is performed within the Aristo memory by choosing from the options presented. When welding, the LCD input screen displays the main welding parameters, average amps, volts, and wire-feed speed.

Other brands of modern pulse-arc MIG welders may have different ways to input the needed parameters but accomplish the same objective.

For example, here are the steps to follow when using an ESAB Aristo:

1. Select between MIG, stick, DC TIG, and Arc Gouging. The MIG process is selected (Figure 6.34).
2. Once MIG has been selected, several options exist within MIG welding, namely spray arc, short arc, pulse arc, and a variation of pulse arc. Pulse arc is selected (Figure 6.35). Scrolling down the Menu the material being welded is the next option to choose.
3. Selection options include steel, stainless steel, Duplex stainless steel, and either 5000- or 4000-series aluminum. Type 5000-series aluminum is selected (Figure 6.36).
4. Choose the wire size. The option .047-inch diameter is selected (Figure 6.37).

   Shielding gas selection is the last major variable to be input. With the high cost of helium, 100-percent argon is selected (Figure 6.38).
5. Once all key variables have been

Fig. 6.33. When in use, the Aristo control panel displays normal welding conditions, wire-feed speed, average voltage, and average current. It has a hold feature that allows the last condition used to be seen after welding has stopped. Wire-feed speed can be adjusted to increase or decrease the amount of heat needed to weld a particular joint.

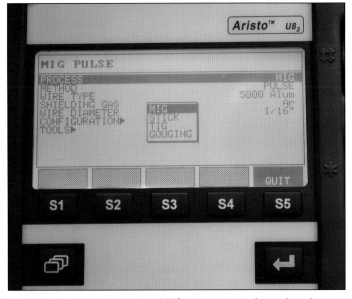

Fig. 6.34. A modern pulse MIG power supply makes it easy to set the required parameters. The Aristo can weld with MIG, DC TIG, stick, and arc gouging. Here, MIG welding is selected.

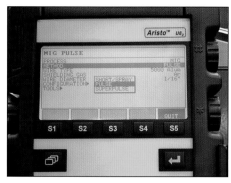

Fig. 6.35. With MIG selected, several options are available: short arc, spray arc, pulse arc, and a special advanced pulse arc. Developing and setting all the required pulse parameters, background current and voltage, peak current, pulse frequency, pulse rise time, and others can be difficult. Modern systems store the information that has already been developed and can be selected with a simple menu-driven approach. Here, pulse-arc MIG is selected.

input (Figure 6.39), the Aristo sets the proper pulse-arc parameters. The memory includes 230 synergic lines for a full range of parameters of wire feed speed that might be used. The welding operator can select and adjust the wire feed speed to match the metal thickness and welding position. The pulse parameters will automatically change to those that will produce sound, quality welds as wire-feed speed is varied.

6. An Autosave feature allows all of the inputs made to be stored in memory for future use (Figure 6.40). Many more features are available, but suffice it to say, that if the application can justify the cost of a microprocessor-based pulse-arc MIG system, it will probably meet your requirements for many years to come.

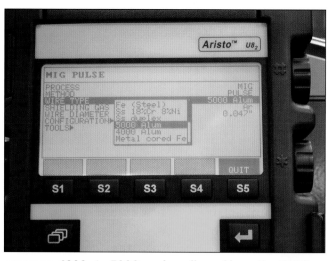

Fig. 6.36. When pulse-arc welding is selected, you need to define the type of material for welding, which determines pulse parameters. You select the specific material type from options built into the Aristo memory. Even for welding aluminum, parameters change from the common 4000- to 5000-series alloys. Here, the 5000-series is selected.

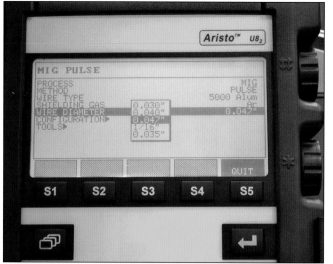

Fig. 6.37. The wire size also defines the pulse parameters. The selection choices available for the 5000-series aluminum alloy are shown. A common size for aluminum (.047, or 3/64 inch) is selected here.

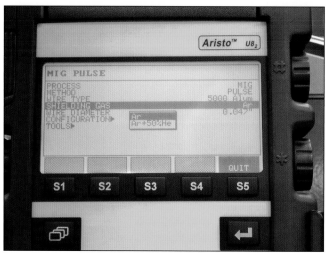

Fig. 6.38. Shielding gas can have a significant effect on the pulse parameters. The choices provided for the 5000-series aluminum are pure argon and a 50/50 argon-helium mixture. A major advantage of pulse-arc MIG is the ability to obtain good wetting and weld penetration with argon. It is not usually necessary to select the more expensive helium gas mixtures (see sidebar "TIG Shielding" on page 45).

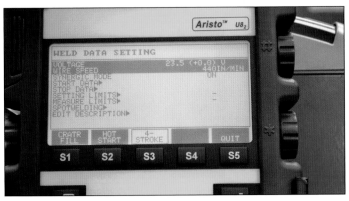

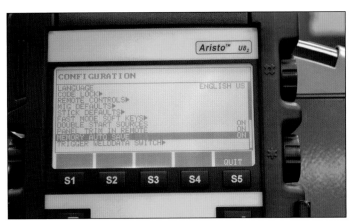

Fig. 6.39. With all the information input, the Aristo selects the proper welding conditions. There are 230 presets built into the memory, which can be accepted or adjusted. There are also many thousands that vary automatically depending on operator input. Synergic Mode is a powerful feature on the third menu line. This allows the weldor to change the wire-feed speed as needed for a specific application. As a result, all pulse parameters change to provide the optimal conditions.

Fig. 6.40. After test welds are made and the needed variables are defined, adjustments can be made to the predetermined parameters. Once satisfied with these changes, these settings can be stored in memory and recalled as needed. Modern welders, like cell phones, have many features that can be accessed and used when needed. Specialized parameters can also be developed and memorized if desired.

Fig. 6.41. Compact portable welders have become very popular. These two provide high-end performance in small, lightweight packages. These Caddy welders can make spray-arc welds within their capacity range and have excellent short arc characteristics. They are both 220-volt input with an output of 160- and 200-amp capacity respectively.

Portable MIG Welders

Two types of portable MIG welders are available: lightweight, professional inverter designs, and conventional heavier transformer designs of lower amperage capacity or duty cycle.

Inverter-based systems: The word inverter is often mentioned when discussing modern advanced welder and plasma cutter designs (see Chapters 3 and 4).

As with stick and TIG welders, for MIG welders, the AC power from the wall power receptacle is first converted to DC, and then the electronic control changes the DC back into AC. That new AC is switching at 20,000 times per second or more, not at 60 times per second as in a common wall supply. This not only allows transformer size to be many times smaller and lighter, it also provides an opportunity to adjust the welding power output thousands of times per second.

Less obvious is the ability to adjust arc current and voltage and current rise times much more precisely than with the large inductor used in conventional MIG welding machines.

Two versions of lightweight inverters are the ESAB Caddy MIG 160i and 200i. The ESAB 200i incorporates a more advanced microprocessor control and LCD readout while the 160i has a simple wire-feed speed knob control.

A MIG welder with inverter features a fast-switching electronic control board (Figure 6.42). With this modern system, a large, heavy, copper and iron inductor is not needed to obtain quality short-arc welds. The electronics on such a welder can look at the amps and volts 20,000 times

Fig. 6.42. Modern welders use fast-switching power devices and electronic circuits to replace iron and copper transformers. The 220-volt 60-cycle AC input line is converted to DC then to high-frequency AC, which allows the main transformer to be many times smaller. Also, a large inductor is not needed because current rise times can be controlled electronically. Even a 200-amp-capacity system weighs only 26 pounds with a built-in wire drive.

Fig. 6.43. This LCD display indicates 159 amps, 21.6 volts, and a wire-feed speed of 350 ipm. On the lower right of the display a material thickness of .11 inch is shown. This is easily set with the lower rotary knob. QSet automatically monitors and adjusts short arc conditions when thin materials are selected.

per second and make adjustments without the need for passive devices, such as inductors.

The ESAB Caddy MIG 200i, without the large inductor, weighs only 26 pounds and has a maximum capacity of 200 amps. It also contains a built-in wire-feed system.

The Caddy MIG 200i has a feature called QSet. It performs several functions. When you input the plate thickness for steel, stainless steel, or aluminum, the amps and volts are preset. You just pull the gun trigger and weld.

In the short-arc mode, QSet looks at the arc for about 6 seconds and adjusts for optimum short-circuiting metal deposition. It retains that setting for subsequent welds until changes are made to the welding parameters. During welding, the LCD display shows the wire-feed speed, amps, and volts.

In Figure 6.44, the material thickness is entered as .15 inch. This is simply changed with the lower rotary knob. Note there is a box around Fe/SS, indicating whether the system is set for steel or stainless steel. Pressing the button directly below the material labels changes the setting to Al for aluminum or CuSi for making a braze weld.

Next to the word QSet is a thermometer symbol. It displays any

adjustments made from those preset by the QSet program.

Figure 6.45 shows the thickness changed from .15 to .03 inch. The preset wire feed speed reduced to 169 ipm.

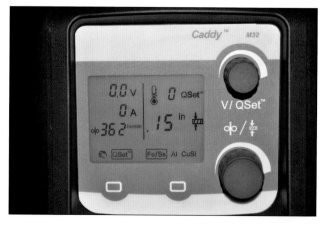

Fig. 6.44. If a thicker base material is selected, the suggested wire-feed speed increases. Here, the material was increased to .15 inch, and the wire-feed speed preset increased to 362 ipm.

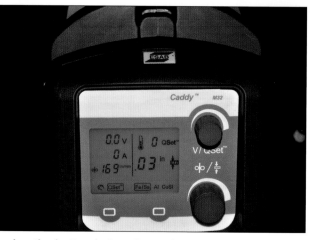

Fig. 6.45. Setting the Caddy 200i to weld the desired material thicknesses is simple. This panel shows three material options: steel, stainless steel, and aluminum. Steel and stainless steel are highlighted in a solid line box. Aluminum and CuSi for (silicon bronze) can also be selected using the button below these three options. Here, the lower right knob has been set at a very thin .03 inch. This defines and sets a wire feed at 169 ipm.

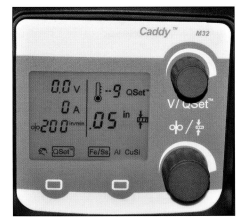

Fig. 6.46. The Caddy 200i sets the proper short-arc condition by monitoring the arc and optimizing current rise time and other parameters for the best arc. If adjustments from the suggested preset values are desired, the upper knob allows adjustment to a colder or hotter setting on a scale of +9 to -9. Here, a colder setting of -9 is indicated. You can see that the indicator on the thermometer icon is at the low end.

Fig. 6.47. If a hotter arc than the preset condition is needed, the QSet thermometer is moved to +5 and the icon is at a higher level. Once set, this higher heat level is maintained as wire-feed speed is increased. The QSet adjusts several parameters to achieve the increase.

In Figure 6.46 the thickness changed to .05 inch and the wire-feed speed increased to 200 ipm. Note that the QSet now reads -9 and the thermometer changed from

Fig. 6.48. Small, non-inverter welders are available inconventional construction design. This Miller unit has a 110-volt input. However, the limited output of 140 amps restricts the metal thickness that can be welded. According to the manufacturer's specification, the maximum thickness of steel that can be welded is 3/16 inch. In addition, this non-inverter welder weighs 60 pounds compared to the 200-amp-capacity Caddy inverter that weighs only 26 pounds. Both include a wire feeder and control.

a mid-setting to no indicator in the symbolic thermometer stem. The QSet preselected welding conditions can be modified from +9 to -9 to fine-tune the settings. The -9 setting provides a colder weld than does the preset middle heat level.

In Figure 6.47, the material thickness remains at .05 inch, but the QSet changed to +5 from -9. Note the thermometer schematic stem now shows the increased black area. QSet adjusts voltage and other parameters to provide a hotter weld than in the middle preset.

Fig. 6.49. This Lincoln welder uses 220-volt input and has a maximum output of 180 amps. It features conventional construction and weighs 66 pounds, but it is rated at 130 amps at 30-percent duty cycle. This means if you weld at 130 amps for 3 minutes, the welder should be allowed to cool for 7 minutes. Be sure to compare maximum capacity with power at various duty cycles when looking at welders to purchase.

Conventional-design portable welders: One example is a Miller 140 (Figure 6.48), which has a 140-amp capacity and weighs 60 pounds. The output capacity is limited because the input power for this welder is common 110 volts AC.

A Millermatic 180 is available that has 220-volt input and 180-amp maximum welding capacity. It weights somewhat more at 72 pounds, but the added output is worth the extra weight if 220-volt input is available.

The Lincoln MIG welder (Figure 6.49) requires 220-volt input, and the latest version weighs 66 pounds with a maximum capacity of 180 amps. If there's a 220 outlet in your shop, purchase a higher-input voltage system, rather than one with a lower-capacity 110-volt input.

Wire Feeding

Whether the wire-feed system is incorporated within the power source or separate, the issues are the same. It is important that the structure supporting the feed rolls and gun inlet be rugged and ridged to maintain wire alignment.

Some weldors use a small-diameter tungsten electrode, a similar size to the wire being used, to ensure proper alignment of the feed rolls and outlet wire guide. If the feed rolls are not aligned, the wire can be easily shaved as it passes through the feeder. The resulting rough edge causes feeding friction and the debris created clogs the gun conduit.

The feed-roll tension control must be properly adjusted regardless of the type of feeder. It should be tightened so the wire feeds at the gun end with a modest restriction. The feed roll adjustment must not be over tightened because that creates metal shavings and the debris clogs the gun conduit liner.

ESAB feed systems have a marked and numbered pressure indicator so the setting can be repeated. If flakes

Fig. 6.50. The quality of the wire-feed system is important. The frame should be rugged so alignment of wire guides, feed rolls, and gun cable inputs are maintained. This heavy-duty unit uses a four-roll-drive system, which is particularly useful for feeding cored wires. A two-roll drive is usually sufficient for solid wire.

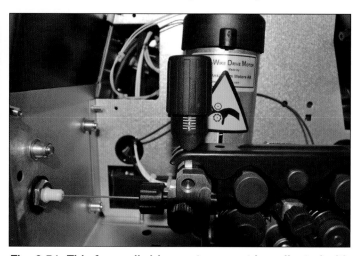

Fig. 6.51. This four-roll-drive system must be adjusted with minimum pressure. With excess pressure, the four rolls can cause wire deformation and roughen the wire surface. That rougher surface causes friction in the gun cable liner and feeding problems.

Fig. 6.52. Setting the top feed-roll pressure is very important, even on two-roll-drive systems. Excess pressure deforming the wire or creating surface shavings is a common problem with feeding wire, and often, metal shavings are found under the feed rolls. Adjust the setting with a clean MIG gun liner and new tip. Only modest feed-roll pressure should be necessary to feed wire as it exits the gun. Use no more pressure than needed.

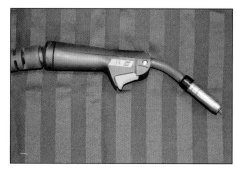

Fig. 6.53. Select a MIG gun with sufficient capacity to match the amount of time spent welding, so it is allowed to properly cool. For example, a 140-amp gun with a 20-percent duty cycle must cool for 8 minutes after 2 minutes of welding.

are observed under the feed rolls, the setting has too much pressure or the feed rolls are misaligned. These flakes look like copper, which makes sense because most steel MIG wire is copper coated. The copper-coated steel flakes are often found to be magnetic.

Stay alert when using two- or four-roll-drive systems. Use just enough tension on the feed-roll adjustment so that the wire does not slip in the feed rolls but not enough tension to distort or put a groove in the wire. If you find filings under the feed rolls, make the required adjustment and clean the conduit liner.

The Gun

The business end of a MIG system is the gun. It has the difficult task of providing a passage for the wire, power to the contact tip, and shielding gas to the diffuser and nozzle. It is also subjected to a great deal of radiant heat from the arc, heat from the contact tip, and resistance heating of the power cable.

Duty Cycle: Weldors want the lightest and longest gun cable possible but that can create feeding problems. For automotive welding, a 200-amp-capacity gun is usually sufficient. Be sure to understand the duty cycle specified in the gun rating. A smaller gun is desirable for getting into small spaces. The gas mixture plays an important role in the amount of heat to which the gun is exposed.

For example, one manufacturer's 160-amp gun is rated at 80-percent duty cycle in carbon dioxide shielding (meaning it can operate for 8 minutes out of 10). That same gun, when used with argon-based gases, is rated at only 100 amps at 80-percent duty cycle. An even heavier 200-amp model is needed for 150-amp capability.

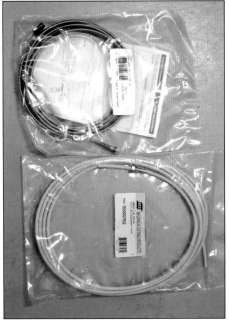

Fig. 6.54. The gun cable liner is an often-overlooked part of the feeding system. For hard solid wire, liners are made of steel. For aluminum, liners are offered in Teflon. When clean, the liner provides a low-friction surface. However, wire-drawing lubricants and shavings at the drive rolls can cause clogging, so liners must be cleaned periodically.

A word of caution: Do not use a MIG gun as a chipping hammer to remove the small amount of slag left on the weld surface or to clean spatter. That usage damages the gun.

Liner: The liner starts at the feed-roll outlet guide and runs through the cable to the contact tip. Refer to the gun manufacturer's instructions to learn how to disassemble and remove the liner for periodic cleaning. Compressed air can be used to blow out debris from the gun end.

A hardened-steel liner wound in a spiral is used to feed steel wire.

A Teflon liner is often offered as an accessory for feeding aluminum wire. If you frequently MIG weld aluminum, consider purchasing a sepa-

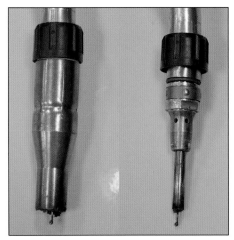

Fig. 6.55. Keeping the gun nozzle clean and free from spatter is important because spatter causes the shielding gas stream to become turbulent and pull air into the shield. Keeping the small holes in the gas diffuser free from spatter is also important. These small holes above the contact tip are carefully placed to allow shielding gas to enter the nozzle, promoting laminar flow. Blocking one or more gas diffuser hole creates a poor gas pattern, and that can cause shielding gas exiting the nozzle to pull air into the gas stream.

rate gun, equip it with a Teflon liner, and do not use it for welding steel or stainless steel. Teflon liners wear very quickly if used for steel wire and keeping a separate gun avoids contamination with debris from steel wire drawing lubricants and copper flaking.

Nozzle: The front end of the MIG gun is where all the action takes place and where problems occur. Figure 6.55 shows a nozzle installed on a gun tip and also removed.

The gas diffuser usually has four to eight carefully angled holes that deliver gas into the gun nozzle. Weld spatter is obvious when it builds up on the gun nozzle. One small piece of spatter can easily clog one or more

Fig. 6.56. When short-arc welding, the contact tip should extend from the gas nozzle as shown on the left, which provides better control of the stickout and delivers better penetration. You are also afforded better visibility with a shorter stickout, so you have improved view of the arc and therefore are better able to keep it on the leading edge of the weld puddle. Some gun manufacturers use longer tips to accomplish the extra extension and some use shorter nozzles. On the right is a standard nozzle.

Fig. 6.57. Nozzles come in various sizes and shapes. For short-arc welding, a 1/2-inch ID nozzle provides adequate shielding gas coverage and better access than a 5/8-inch ID nozzle. The larger sizes are better suited to higher currents or conditions in which a higher flow is needed to counter drafts. Tapered straight-end 1/2-inch nozzles are on the left, and a 3-4-inch nozzle is on the right.

Fig. 6.58. A special nozzle (left) and a standard nozzle (right) are available for MIG spot welding. They are designed to press against the top of a sheet to help provide good contact with the lower sheet. Usually made of thick-wall material, they are designed with vent passages at the bottom to allow shielding gas and spatter to escape.

of these gas-diffuser holes and affect the quality of gas shielding. Be sure to check the gas-diffuser passages every time a contact tip is replaced or nozzle cleaned of spatter.

When short-arc welding, the contact tip should extend outside the nozzle. This ensures the use of the proper short stickout to maintain good penetration. There are two ways to accomplish this—longer contact tips or shorter nozzles.

Nozzles are available in a number of different diameters. Some manufacturers use one system, where a number-8 nozzle is 8/16 inch long and has an inner diameter (ID) of .50 inch. Others list them by ID size in inches, i.e., 3/8, 1/2, 5/8, or 3/4. For most sheet-metal and thinner material welding, a 1/2-inch ID nozzle is the largest size needed. At times, for greater access to the weld joint, a 3/8-inch ID nozzle is preferred.

A spot-welding nozzle can be useful when making spot and tack welds in sheet metal. This nozzle is longer than a standard nozzle and can be pressed against the workpiece while providing the proper wire stickout. It has openings at the bottom to allow shielding gas and spatter to exit. It usually has thicker walls so it can handle increased heat from arc radiation. The thicker walls also help provide structural strength

so it can be pushed against the top sheet onto the bottom sheet. A spot weld timer built into some welders can help ensure the proper amount of weld penetration without burning through.

Shielding Gas Flow Control

There is often confusion about how MIG gas-flow controls operate and what amount of flow is needed. With MIG (and TIG), welding flow rate is measured in cubic feet of gas

Fig. 6.59. One type of shielding gas control system employs a fixed-pressure regulator followed by a flow tube that has a ball, which rises when welding. It reads gas flow in CFH. Following the flow tube is an adjustable-needle valve that reduces the pressure from the regulator to that needed to flow the desired amount of shielding gas. The typical pressure in the gas delivery hose when welding varies from 3 to 8 psi.

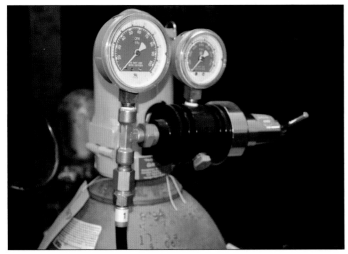

Fig. 6.60. *The other common type of shielding gas control system uses an adjustable regulator and a small-diameter orifice. The output-gas connection fitting on the regulator has a very small hole—as small as 0.025 inch. The orifice drops the pressure to that needed to flow the desired amount of shielding gas. These devices are designed to operate in a mode called choked flow, so only the upstream pressure controls flow.*

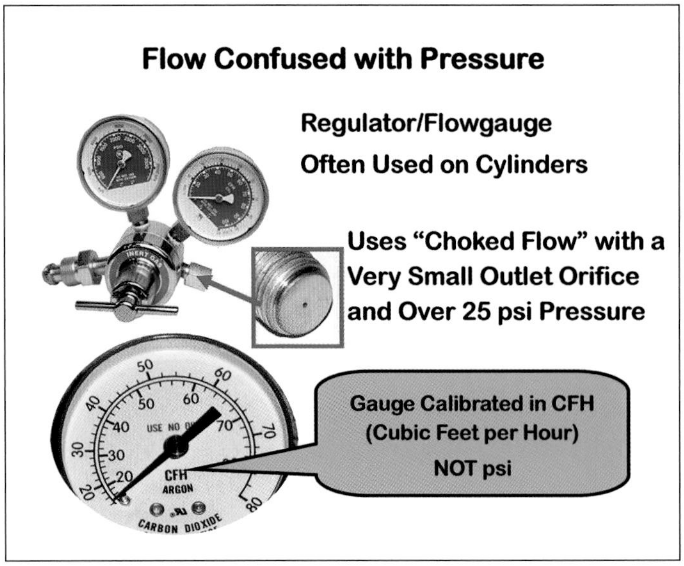

Flow Confused with Pressure

Regulator/Flowgauge
Often Used on Cylinders

Uses "Choked Flow" with a
Very Small Outlet Orifice
and Over 25 psi Pressure

Gauge Calibrated in CFH
(Cubic Feet per Hour)
NOT psi

Fig. 6.61. *Regulator flow gauges can lead to confusion about what is being set—flow or pressure. For MIG welding, flow rate is set (not pressure). Note the very small orifice on the regulator output. It can be as small as .025 inch. Also note the output gauge is labeled as CFH, not PSI. Most of these regulators operate at 40 to 60 psi.*

per hour, not pressure, at least not directly. There are two types of flow control regulators: one uses a flow meter and the other uses a flow gauge.

A regulator flowmeter is often used in MIG and TIG welding, and uses a regulator to set a fixed pressure from 25 to 80 psi.

A flow tube has a ball that rises when gas is flowing. The height of the ball accurately defines the shielding-gas flow rate in CFH. Flow tubes are calibrated at the preset pressure of the regulator, such as 25, 50, or 80 psi.

An adjustable needle valve drops the pressure to the proper level for flowing the specified rate of shielding gas through the MIG gun and gun cable. The gun usually requires shielding gas pressure from 3 to 8 psi, but the pressure varies as the gun cable is bent and looped and as spatter accumulates in the nozzle.

A regulator flow gauge sets pressure above a very small orifice usually located in the regulator gas outlet fitting. The orifice can be as small as .025 inch. This allows the gas flow to be varied by adjusting the regulator pressure.

The model shown in Figure 6.60 is for welding with 100-percent carbon dioxide. It has a heavy, finned, black body to absorb heat because high-carbon-dioxide flows can cause ice to form in the flow-control orifice, reducing flow. The orifice in the flow gauge drops the pressure set by the regulator to the 3 to 8 psi needed to flow the desired amount of shielding gas.

Many MIG systems are sold with regulator flow gauges, causing some of the confusion about setting pressure versus setting flow. The output

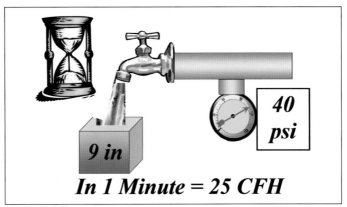

40 psi

9 in

In 1 Minute = 25 CFH

Fig. 6.62. *This analogy may help explain the difference between flow and pressure. A water pipe in a house typically operates at 40-psi pressure. If the valve is opened slightly, the flow fills a 9 x 9-inch box in 1 minute if it flows about 25 cfh (which is similar to shielding gas flow rate in a MIG system). The pressure in the pipe ahead of the faucet valve remains at 40 psi at these low flows.*

MIG GUN NOZZLE SIZE INSIDE DIAMETER	MINIMUM SUGGESTED FLOW	TYPICAL FLOW SETTING	MAXIMUM FLOW SETTING
3/8 INCH	15 CFH	18-22 CFH	~30 CFH
1/2 INCH	18 CFH	22-27 CFH	~40 CFH
5/8 INCH	22 CFH	30-35 CFH	~55 CFH
3/4 INCH	30 CFH	30-40 CFH	~65 CFH

Fig. 6.63. This chart shows suggested MIG gas flows for various-size nozzles. Note the maximums as well as the minimums. If the maximum is exceeded, the shielding gas stream becomes turbulent and pulls in air, which is counterproductive. If there are drafts present, use a windbreak but do not exceed these flow levels.

gauge is not labeled by psi as are normal pressure gauges. It is calibrated in CFH. Knowing the orifice size, the flow-through can be determined, depending on the regulator output-pressure setting.

The water analogy in Figure 6.62 may help explain the flow-versus-pressure issue. In the water pipe leading to a house faucet, the pressure remains at 40 psi while the valve is slightly opened and water flows at a low rate. If the flowing water fills a 9-inch box in 1 minute, it is flowing 25 cfh, about the same rate of shielding gas flow used while welding. The flow rate can be varied in this low range by turning the faucet handle and the pipe pressure remains at 40 psi.

The chart in Figure 6.63 shows the minimum, typical, and maximum flow rates that should be used when MIG welding. If the maximums are exceeded, a turbulent gas stream is created that pulls air and any moisture present into the shielding gas.

I have measured flows in excess of 200 cfh at the weld start that lasts for several seconds. It also takes time for the gas stream to stabilize to the desired smooth flow after the flow rate is returned to normal. The high gas pressure that builds in the hose when welding stops causes this blast of gas at the start of the next weld.

Why is the high pressure needed? Restrictions occur in the gas flow system due to spatter build-up in the nozzle and gas diffuser as well as in the gun cable conduit where the gas passage often doubles as the hose that holds the spiral wire liner. If a fixed, lower pressure were used, the restrictions would change the preset flow.

The engineers that designed MIG gas flow systems knew flow restriction

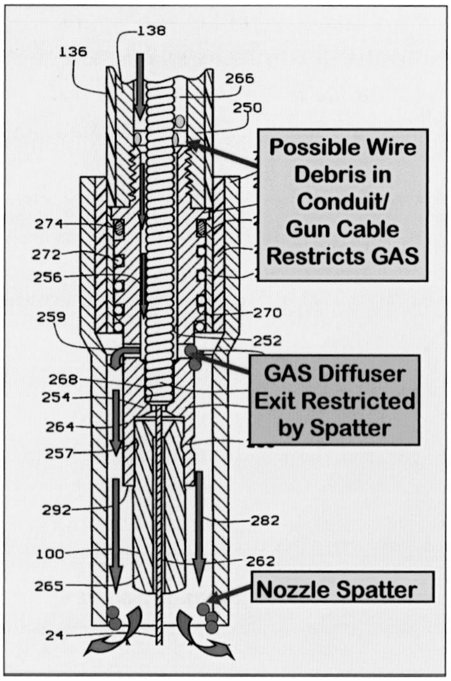

Possible Wire Debris in Conduit/ Gun Cable Restricts GAS

GAS Diffuser Exit Restricted by Spatter

Nozzle Spatter

Fig. 6.64. Flow restriction occur when spatter builds up in the gun nozzle or blocks the gas diffuser holes. When the gas passage in the gun cable is bent, looped, or clogged with debris, it creates a less obvious source of restriction. In fact, the gas hose in many guns doubles as the passage holding the spiral wire liner. Copper flakes and drawing lubricant can build up over time and cause restrictions.

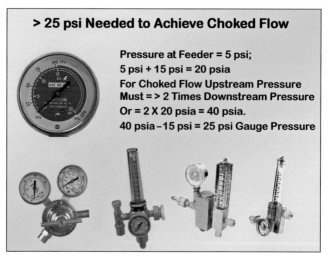

> 25 psi Needed to Achieve Choked Flow

Pressure at Feeder = 5 psi;
5 psi + 15 psi = 20 psia
For Choked Flow Upstream Pressure
Must = > 2 Times Downstream Pressure
Or = 2 X 20 psia = 40 psia.
40 psia – 15 psi = 25 psi Gauge Pressure

Fig. 6.65. Choked flow occurs when the absolute pressure (gauge pressure + 14.7 psi) above the flow-control needle valve or orifice is twice the downstream absolute pressure. Quality flow controls for MIG are maintained from 25 to 80 psi, all above the 25 psi needed to achieve choked flow.

The use of high pressure solves the flow consistency problem, but high-pressure gas flows into the gas delivery hose every time welding stops. As a result, when welding resumes, excess gas stored in the hose rushes out. Much of the gas is wasted and air is pulled into the gas stream.

variations occur while welding. They used a principle called choked flow to ensure that flow rate does not change when these restrictions occur.

To achieve choked flow the absolute pressure upstream of the flow control needle valve or orifice must be twice the absolute downstream pressure. Absolute pressure is the pressure gauge reading plus atmospheric pressure (15 psi for these calculations.) A minimum of 25 psi is needed (see Figure 6.65) for quality regulator flowmeters. Regulator flow gauges typically use pressures from 40 to 60 psi.

Why is that important? Because every time welding stops or the wire is inched forward to cut off the end, the pressure in the gas hose from the cylinder to the welder quickly rises to the regulator pressure. This increase in pressure causes excess gas to be stored in the gas delivery hose until the next weld starts and excess gas blasts out of the gun nozzle. Typically, more than

half the total shielding gas used by most MIG welders is wasted because of this starting blast of gas.

A way to control this excess gas blast at the weld start is to reduce the volume of gas stored in the hose when welding stops. Figure 6.66 shows the special gray gas delivery hose attached to the output of the regulator flow gauge. It is a patented device that simply replaces the existing gas-delivery hose. It holds only about 20 percent of the volume of the typical 1/4-inch ID gas-delivery hose that comes with many MIG welders. This simple device can cut total shielding gas use in half or more (see NetWelding.com for details).

I use a large-capacity shielding gas cylinder in my own workshop. It is much too large to place on a small rolling cart so it is chained to the wall. It has a 25-foot gas hose from the cylinder to the welder. That long length would waste considerable shielding gas at each start if it were equipped with a standard-diameter hose. The resulting gas blast at the weld start would also pull air into the

shielding gas stream creating excess spatter and inferior weld-start quality. The custom-extruded, small-ID, thick-wall gas hose saves gas and improves weld starts. In addition to the small ID, a peak-flow limiting orifice is incorporated in the gas hose fitting on the welder end of the hose.

Purchasing a MIG System

Many options are available, which makes the choice more difficult. Certainly your budget is a big factor. Systems that work but have limited capability can be purchased for a few hundred dollars from companies such as Harbor Freight. They may have limited capability, such as only usable with self-shielding flux-cored wire, which has no provision for using a shielding gas. These lower-capacity welders usually operate at a limited-duty cycle. For example, if labeled for 20 percent and the welder is operated for 2 minutes, you need to wait 10 minutes for the internal power components to cool off. With certain welders, I have also found

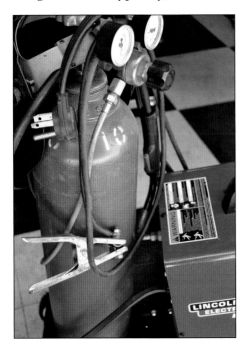

Fig. 6.66. This welder uses a gas-delivery hose with only 20 percent of the volume of a typical gas delivery hose. It reduces the gas surge by 80 percent.

Fig. 6.67. In my shop, a large-capacity shielding gas cylinder is chained to the wall and the gas delivery hose is 25 feet from the welder. Significant gas waste would exist if it were not for the use of this custom, extruded, gas-saving delivery hose. It simply replaced the standard large-diameter gas hose.

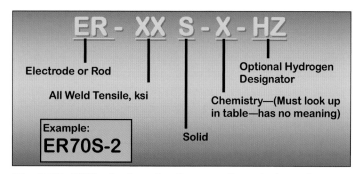

Fig. 6.68. AWS wire type is often mentioned when choosing a steel wire or rod. This illustration shows what the letters and numbers mean.

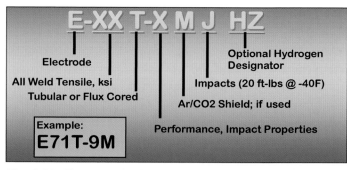

Fig. 6.69. Flux-cored wires have a similar AWS designation as solid wires. This illustration shows what the letters and numbers mean.

you need to wait even longer if you are using the welder on a larger welding job, as it may take even longer to cool properly the second, third, etc. time you pull the gun trigger. If your budget only allows this purchase amount, you will no doubt want to upgrade in the future to have a more professional MIG welder.

If you can spend somewhat more, MIG welders made by the leading welding companies are sold at Home Depot, Lowes, Tractor Supply, and other locations. Many have the ability to weld with shielding gas and incorporate a gas solenoid and include a MIG gun that can provide adequate gas shielding. Most can also weld without gas using self-shielding flux-cored wires. Both self-shielded and sold wire requiring shielding gas are also sold at these retailers. I recommend saving up and purchasing this quality-level welder.

In addition, if you have 220-volt input power available in your workshop, I recommend that you purchase the highest current rated welder that the higher voltage input allows. It can also operate at low currents and has the capability of welding with larger wire diameters, such as .045 inch, to obtain good penetration on material 1/4 inch and thicker.

If you're planning to use gas

All Wires Below Require Minimum: 70 ksi Tensile Strength, 22% Elongation and Impacts of 20 ft-lbs @ -20 F				
AWS Alloy Designation	C carbon	Mn manganese	Si silicon	Comment
ER70S-2	.07 max	.90 to 1.40	.40 to .70	Most Ductile; less fluid puddle for all position welding
ER70S-3	.06 to .15	.90 to 1.40	.45 to .75	Common lower cost; for clean, low scale surface
ER70S-6	.06 to .15	1.40 to 1.85	.80 to 1.15	Hi Si content; handles more scale
ER70S-7	.07 to .15	1.50 to 2.00	.50 to .80	Better Mn/Si ratio assures better wetting

Fig. 6.70. To fit all products into AWS materials standards can be difficult and may cause the chemical ranges to be very broad. That is the case with ER70S-6, a commonly used MIG wire. If an S-6 wire is on the high end of silicon and low end of manganese, the wetting can be less than desirable. The use of ER70S-7 avoids that possibility.

shielding, check with the local gas suppliers for the cost of cylinder gas and the sizes of cylinders they sell. The largest industrial-size cylinders are usually only rented, not sold. Unless you're doing production welding, these are usually larger than needed. While at the gas distributor, find out what size welders they offer. If you can afford the extra money they will no doubt have a 200- or 250-amp MIG welder that may come with wheels and can hold a gas cylinder on the back. These welders usually have duty cycles in the range of 60 percent, which means that in a 10-minute period they can weld for

6 minutes before needing to cool for 4 minutes. These typically are properly rated and can weld 6 out of 10 minutes all day long. In fact, it may be even a higher duty cycle if you are not using the full-rated current capacity. These industrial welders often can produce a better weld quality with less spatter and better starts.

The choice is yours and what your budget allows.

Wire and Gas Selection

You can MIG weld several types of material including steel, stainless steel, and aluminum. Most material

MIG Filler Specifications

An interesting experience related to ER70S-7 wire occurred in our lab. A prominent fabricator was having weld-wetting problems when joining pipe with MIG short-arc welding using ER70S-6 wire. One of the best weldors I have ever known, Chuck Hradil, visited the fabricator to investigate the problem.

Because of Chuck's considerable experience and talent, he was expected to make good welds under almost any circumstance. He came back from the customer visit and sat in my office very upset because, after spending several days trying, he could not make quality welds. They were too convex and not well wet into the edges.

The fabricator's setup was duplicated in our lab and the particular heat of ER70S-6 wire was tested. Although the wire chemistry was within the AWS specification, it was on the very high end of silicon content and was low in manganese. It had a ratio close to the allowable AWS minimum of 1.2:1.

At the time, a special MIG wire alloy was being made for a farm equipment manufacturer according to its specification, but it did not have an AWS classification. The pipe welding fabricator switched to this special alloy, and the problem was resolved and never reoccurred.

That wire's chemistry range forced a minimum ratio of about 2:1 or higher. AWS provided a classification for the alloy, and when approved, it was designated ER70S-7. ■

Process Variation	Typical Gas Mixtures for Welding Carbon Steel	
Spray Arc	2 to 5% O_2 in Argon	5 to 10% CO_2 in Argon 8% CO_2; 2% O_2; 90% Argon
Short Arc	15 to 25% CO_2 in Argon	8% CO_2; 2% O_2; 90% Argon
Pulse Arc	2 to 5% O_2 in Argon	5 to 8% CO_2 in Argon

Fig. 6.71. For carbon steel, the shielding gas used depends on the process variation employed. Quality spray-arc can be performed with 2- to 5-percent oxygen in argon; 5- to 10-percent carbon dioxide is also common. Optimum short-arc metal transfer requires a minimum of 15-percent carbon dioxide with a very common 25-percent carbon dioxide used by a many weldors. Note the triple mix of carbon dioxide, oxygen, and argon can be used for both short arc and spray arc. If investing in pulse-arc equipment, the optimum gas mixture should be used.

used for street rods and race cars is made of steel. Much of the steel used is carbon steel, not high-strength steel (discussed in Chapter 4).

Steel

The AWS defines the standards that MIG wires must meet to be labeled on the product package with their AWS classifications. Specific chemistry and weld property requirements must be achieved after making a weld test with a representative sample of the product.

See Figure 6.68 for a breakdown of the numbers and letters in the AWS classification system for steel wires.

Many of the wires listed for steel welding have similar minimum tensile and impact property requirements. The biggest difference among them is wire chemistry. For solid wire, it is the number following the letter S such as S-2, S-6, and S-7. The specification must be referenced to define the specific chemical requirements.

There is also a standard for gas shielding and self-shielding steel flux-cored wires, which includes metal-cored wires (see Figure 6.69).

The most important difference between MIG solid-steel wires is the required range of chemistry. The number following the S in the description

defines the allowable range. (There is no significance to the actual number, a higher number does not mean a better wire. It often reflects the time period the alloy was added.)

Although there are some unique benefits to using an ER70S-7 wire, they are not because it has a higher number. Figure 6.70 shows most of the wire alloys used in automotive welding of steel. They all have about the same strength and toughness.

An important factor is the manganese-to-silicon ratio. Ideally it should be greater than about 2:1.

Weld wetting is improved when the ratio is about 2:1 or higher. With a commonly used ER70S-6 alloy, the allowable range can cause the manganese to be as low as 1.40 and the silicon to be as high as 1.15. That gives a ratio of 1.40/1.15 or 1.2:1. This creates a weld that does not wet as well at the edges and may cause undercut.

The lowest ratio that can occur with an ER70S-7 is: minimum manganese of 1.5 and maximum silicon of .8, which gives a ratio of 1.9:1.

Most of the time the ER70S-6 may be in an acceptable range but with ER70S-7 it is ensured.

The shielding gas used when MIG welding steel has a significant effect on weld appearance, quality, and the amount of spatter. It is possible to weld with 100-percent carbon dioxide. The metal transfer is globular. It does not produce spray-arc transfer, and it cannot be used for pulse arc welding. However, it does allow the production of reasonably quality short-arc welds.

Welds produced with 100-percent carbon dioxide are also hotter and should be limited to metal thicknesses greater than 1/8 inch. The spatter levels are higher than argon-based gas mixtures, requiring more post-weld cleaning.

For example, at a typical 150 amps, carbon dioxide produces about 6-percent spatter compared to welding with a typical argon-based, 25-percent carbon dioxide mixture where spatter is about 3 percent. Because sheet metal is often welded and low spatter is a significant benefit to reduce post-weld cleanup, the use of argon-based gas is recommended.

However, there have been significant recent increases in the price of argon and argon-based shielding gases due to its limited availability. Carbon-dioxide shielding may be satisfactory for your applications and is a more economical choice. Argon is only .9 percent of the air and is produced by liquefying air then distilling the oxygen, nitrogen, and small percentage of argon. However, producing argon requires additional expensive equipment and a more elaborate production process. It is economical to produce only if the majority of 98 percent of the air, the oxygen and nitrogen, can be sold.

Carbon dioxide can be produced by a number of methods, including burning hydrocarbon fuel. However, it is often produced as a byproduct of other chemical production. In addition, similar-size gas cylinders contain about 1.8 times the amount of carbon-dioxide gas as argon because carbon dioxide becomes a liquid when pressurized. This means fewer cylinder changes for the user. In addition, the cost of transporting both gases is the heavy steel cylinders required. About 70 to 80 percent of the weight of a full cylinder is the steel. Therefore the transportation cost per cubic foot of carbon dioxide gas is 1.5 to 1.8 times less depending on the specific cylinder. Check on the local price of cylinders in your area.

Pure argon cannot, by itself, be used for welding steel. To obtain good arc stability and low spatter for short-arc welding, some amount of carbon dioxide or oxygen is necessary in the gas mixture. The typical mixtures include 25-percent carbon dioxide and 75-percent argon. A mixture with 90-percent argon, 8-percent carbon dioxide, and 2-percent oxygen can be used for both spray arc and short arc.

Stainless Steel

Welding stainless steel is similar to welding steel. Wire and gas selection are different, but feed rates and welding procedures are similar. Stainless-steel welding wire, like steel wire, is stiff, and the same feed systems are used.

MIG welding of stainless steel differs from what has been covered in TIG welding. To obtain well-wet edges, some oxygen and carbon dioxide must be added to the shielding gas. The allowable amount of carbon dioxide must be limited to maintain corrosion-resistant properties.

Several welding-wire alloys are available, and in general, the wire numbers match the plate material designation. The exception is 308 welding wire that is used on most 18-percent chrome and 8-percent nickel alloys, such as the most common plate, type 304. For MIG welding, a low-carbon version of 308 (designated 308L) reduces the problem of chrome carbide forming in the grain boundaries. If excess carbon is present, it combines with chrome, depleting the effective chrome in the steel. Chrome is the main element in stainless steel that provides corrosion resistance. The addition of silicon to the wire chemistry helps improve edge wetting. A good wire choice for welding most stainless steels is designated 308LSi, which has low carbon with added silicon.

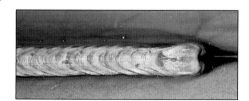

Fig. 6.72. *To effectively weld stainless steel, the appropriate shielding gas mixture and the proper welding wire chemistry must be selected. The shielding gas selection affects the quality of the finish and edge wetting. Stainless-steel wire is stiff and its feeding issues are similar to those with steel.*

Typical Stainless Steel Welding Wire Alloys

Wire Alloy	Cr	Ni	Mn	Other	UTS, ksi	Yield, ksi
308LSi*	20%	10%	2%	.8% Si; .02 C	90	60
312**	30%	10%	1.8%	1.8%	85	55
OK 6.95 ***	18%	8%	7%		92	65
316L	19%	12%	1.6%	2.4% Mo	83	60

Note*   For welding 304 and most typical stainless base materials.
Note**   Alloy 312 often used for welding stainless to steel.
Note*** Alloy OK 6.95 does not have an AWS class. It can be.
                used for welding 409 stainless muffler material.

Fig. 6.73. A number of stainless steels are available, but 18-percent chrome and 8-percent nickel is commonly used for automotive applications. Type 304 is the most common, and it is welded with 308 wire. For MIG welding, the use of a low-carbon wire with added silicon is suggested. For more corrosion resistance, such as in marine use, 316 stainless steel (with the addition of moly) is commonly used.

Stainless Steel MIG Shielding Gases

Process	1st Gas Choice	2nd Gas Choice
MIG Spray Arc	(1) 98% Ar, 2% O_2	(1) 99% Ar, 1% O_2
MIG Short Arc (2)	90% He, 7.5% Ar, 3.5% CO_2	(3) 97% Ar, 2% CO_2, 1% H_2
MIG Pulse Arc	(2) 95% Ar, 5% CO_2	(3) 95% Ar, 3% CO_2, 2% H_2

Note (1) The surface will be oxidized and require brushing.
Note (2) Although 8% CO_2 is usually considered excessive if limited to
                single pass Short Arc welds it may be satisfactory. See text.
Note (3) For welding 304 and most 18% Cr / 8% Ni stainless materials.

Fig. 6.74. MIG stainless-steel shielding gas is different from what is commonly used for steel. It is usual to keep the carbon dioxide below about 5 percent for good corrosion properties. However, I use a mixture usable for both steel and short arc stainless steel: 90-percent argon, 8-percent carbon dioxide, 2-percent oxygen. The use of a low-carbon wire when welding stainless steel, such as 308L or 308LSi, is suggested if that mixture is used.

MIG stainless-steel shielding gas is different from what is commonly used for steel. One of the most common shielding gas mixtures (75-percent argon and 25-percent carbon dioxide) used for welding steel has excessive carbon dioxide for welding stainless steel. As mentioned, carbon combines with chrome at the grain boundaries and reduces the effective chrome in the material. When carbon dioxide is present in the shielding gas, carbon is absorbed into the molten weld metal.

It is usual to keep the carbon dioxide below about 5 percent when welding stainless steel. Therefore, a stainless-steel gas mixture is needed. For MIG short-arc welding, which is most often used for automotive applications, the recommended gas mixture is a special tri-mix containing 90-percent helium. Because of the high cost of helium, alternatives should be considered.

Note the mixture recommended for pulse arc (5-percent carbon diox-ide in argon) may provide acceptable results when welding sheet metal. In heavier thicknesses (greater than 3/16 inch), it may not provide sufficient penetration in the short-arc mode.

I use a MIG gas mixture of 8-percent carbon dioxide, 2-percent oxygen, and 90-percent argon in my workshop. It performs well for steel and for some stainless-steel applications. It has slightly more carbon dioxide than is recommended for welding stainless steel, but for single-pass short-arc welds it is satisfactory.

Carbon pick-up from the shielding gas is less when using short-arc than when using spray-arc welding. A low-carbon wire, usually desig-nated with the letter L following the alloy designation (308L or 308LSi), is suggested if that mixture is used. One important benefit of this gas mixture is that a different gas mixture and cylinder is not necessary for welding steel and stainless steel. If welding stainless-steel headers where the temperatures can get quite high, TIG welding with 100-percent argon is the best choice to avoid the corrosion issue all together.

Aluminum

MIG welding aluminum has a number of differences from welding steel or stainless steel. An oxide forms rapidly on aluminum that has

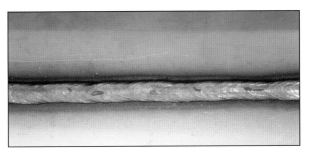

Fig. 6.75. This fillet weld was made with MIG short arc metal transfer. A few slag islands formed on the surface and caused some discoloration, but it can be wire brushed to make the appearance uniform.

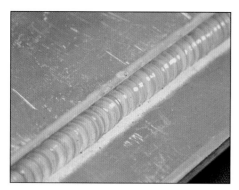

Fig. 6.76. MIG welding aluminum presents challenges but can produce excellent weld results. This weld has two light areas next to the weld. These cleaned areas were produced while welding and are important to achieving good weld quality.

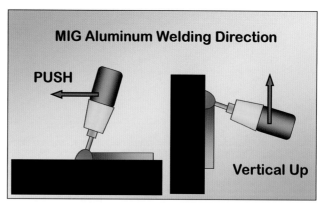

Fig. 6.77. There are a number of important differences between MIG welding aluminum and steel. One is the torch angle. It is necessary to weld with a push (not pull) angle. Upward motion must be used when welding in the vertical position.

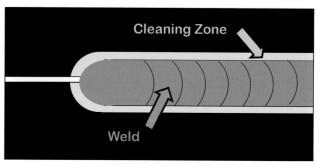

Fig. 6.78. Aluminum welding rapidly forms a high-melting-point oxide, and that prevents molten metal from wetting into the surface. Aluminum welding is performed with DCRP power. The electrons leaving the base metal and shielding gas atoms hitting the surface break up the oxide and leave a cleaned area in front of and next to the weld.

a very high melting point, and the molten weld pool does not displace it. It must be removed just before welding, or the weld does not wet into the base material. In addition, any contact with iron or iron oxide creates poor welds.

Therefore, saw blades, files, sanding disks, and wire brushes must be dedicated for use on aluminum and not be contaminated with steel. A stainless-steel (not steel) wire brush dedicated for use in aluminum welding is an important tool. The surfaces to be welded should be brushed so that the surface oxide is removed.

Gun Angle: As with TIG welding (see Chapter 4), when the wire is positive polarity there is a cleaning action that helps remove the oxide formation on the plate surface in front of the arc. To maximize this effect, use a push welding position. For vertical welds, the weld should be made in an upward motion, which is also preferred for welding steel.

Cleaning Zone: Figure 6.78 shows the cleaning zone created by the action of electrons leaving the plate surface on way to the welding-wire

tip and the positive-charged gas ions traveling in the opposite direction, striking the plate surface.

The cleaning zone is clearly seen in the fillet weld in Figure 6.79. However, cleaning the base material cannot be left only to the power source. The surfaces must be free from all oil or grease and cleaned with a solvent if necessary.

Then a dedicated stainless steel brush should be used to remove the oxide layer just before welding. Files can be used on the mating surfaces of the joint, but as with wire brushes, they should only be used for cleaning aluminum and not contaminated with steel.

Wire: Feeding of the soft aluminum wire often causes frustration for skilled steel MIG weldors. A bulky spool gun that uses expensive small spools of wire is often the recommended solution.

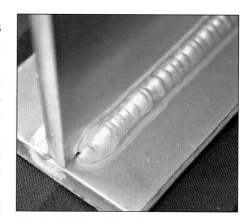

Fig. 6.79. Here, the cleaning zone is in front of the aluminum fillet weld. Cleaning the surface oxide is very important when welding aluminum to achieve proper weld wetting. The plates to be welded must also be cleaned of the surface oxide just before welding.

A push MIG gun system can be used for welding aluminum if careful attention to feeding details is taken. First, if possible, avoid using the

Fig. 6.80. Most weldors have difficulty feeding MIG aluminum wire. Using a conventional MIG weld system in which the wire is pushed through a MIG gun is possible but takes special attention. Quality aluminum wire is thread-wound on the spool and, with each turn, the wire is stacked neatly, one strand next to the other. The tension holding the spool on the take-off spindle prevents unwinding, but it does not have to be as high as with steel. Set the minimum amount.

Fig. 6.81. Special aluminum-wire feed rolls are available for most feeders. On the left is a U-groove feed roll designed for feeding aluminum. In the middle is a feed roll with a V-angle design, which is supplied with most MIG welders. On the right is a feed roll for flux-cored wire that should not be used, even on solid steel wire.

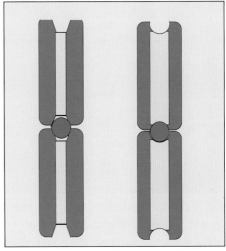

Fig. 6.82. This illustration shows the difference between V- and U-groove feed rolls. The V-angle system works fine for hard steel wires where the wire contacts the feed rolls at four points and can supply griping pressure to feed the wire. For aluminum wire, that small contact area can scratch the wire and make small grooves which create friction. The smooth U-groove cradles the wire and feeds without marring the surface.

soft 4043 welding wire for the structural alloys often used for automotive applications. In most instances, stiffer 5356 wire is acceptable.

Be sure to buy a good-quality and reliable brand of welding wire. A surface finish on the wire can make feeding easier. Aluminum should be thread wound on the spool, and each turn neatly placed alongside the next. Use the minimum tension on the device that prevents the spool from turning on the spindle to reduce the amount of load needed to pull the wire from the spool. Aluminum, unlike steel, does not unwind as readily and lower spool tension can be used.

Feed Roll: Another important consideration is the feed roll. Purchase

feed rolls specifically for aluminum. Typical types are shown in Figure 6.81. The one on the right is serrated, used for feeding flux-cored wires. The one in the center is typical for feeding steel wire. It has V grooves, which is fine for feeding steel but not for feeding aluminum wire. The feed roll on the left has U grooves and should be used when feeding aluminum.

The differences between V-groove and U-groove feed rolls may seem

subtle, but they are very significant (see Figure 6.82). The V groove only contacts the steel wire at four points on the sides of the upper and lower V. This allows it to handle variations in

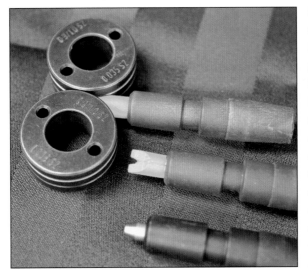

Fig. 6.83. The outlet guide can help with aluminum wire feeding. This device guides the wire from the feed rolls to the gun-cable wire inlet. For welding with steel wire, it is typically made from steel. For aluminum, a tapered plastic outlet guide is used (yellow), so it can be placed closer to the feed rolls and not damage the wire. However, it wears and needs to be periodically replaced.

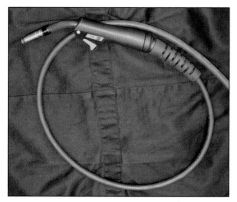

Fig. 6.84. To weld aluminum, you can use the same MIG gun as used for steel welding, but it must be equipped with aluminum accessories. When welding aluminum, never use a contact tip that was used for steel. Keep a special tip for aluminum use only. If the gun and cable being used for steel is longer than 10 feet, buy a new one that is a maximum 10 feet long and keep it for use with aluminum.

diameter that occur in wire manufacture and still provide good feed force. The steel wire does not deform, and the gripping force remains over the range of wire diameter tolerances.

Aluminum wire is too soft for feeding with a V-feed roll system. The wire distorts and may cause a rough surface, which increases friction in the gun cable and the gun itself. The U-groove feed rolls essentially cradle the wire, and their highly polished surface do not cause surface roughness.

Outlet Guide: This supports the wire as it leaves the feed rolls. On the bottom of Figure 6.83 is the typical outlet guide used for steel. It has a

Fig. 6.85. The gun cable liner is the main change that's needed for feeding aluminum. While a spiral steel liner is used for welding steel, a Teflon or plastic liner is necessary for aluminum. Follow the gun manufacturer's instructions and carefully measure and adjust the liner. It may be easier to have a separate aluminum welding gun, rather than switching between steel and Teflon liners.

cylindrical steel tip that comes close to the feed roll. The yellow plastic outlet guide is designed for feeding aluminum wire. It has a tapered end and can contact the feed roll. It supports the wire without a gap. It cannot be used for feeding steel as it would wear very quickly.

As mentioned previously, you must set the pressure on the feed-roll system to provide just enough force to feed the wire. It must be adjusted when welding because the forces

increase when the arc is established and the contact tip gets hot.

Liner

A key item in achieving successful feeding of aluminum is the MIG gun and liner. If welding aluminum frequently, it may be beneficial to purchase another gun. It is much less expensive than purchasing a bulky spool gun with its separate feeder and control. The gun can be set up specifically for aluminum for each use. If the MIG gun cable used for welding steel is greater than 10 feet long, purchase a shorter one for aluminum. Also, when welding, position the welder so that the cable is as straight as possible.

Use a Teflon liner for feeding aluminum. Some gun manufacturers offer both a Teflon liner and a less-expensive plastic liner. Both can work, but Teflon provides less friction. Be sure the contact tip is not too tight for the size of wire.

Projects and Applications

The MIG process is widely applied to a broad range of applications from thin automotive-type sheet metal to the welding of very heavy plates used on ships. It has a number of advantages, and most weldors can learn to make quality weld deposits with practice. The following specific applications will help you understand why the process is preferred for many applications.

Application: Creating an Automatic Gate Latch

For this application a small inexpensive MIG welder could be used with self-shielded wire. In fact, this is an outdoor application so it is preferred over gas shielding to avoid

wind affecting the shielding gas, which would cause poor weld quality.

Several years after having a two-section 10-foot-wide iron gate installed at the entrance to the rear

of our house, the home insurance company required that the gate be self-closing with a latch that was sufficiently high off the ground so it could not be reached by a small

Fig. 6.86. Our home insurance company required that the gate to the back yard close and latch automatically. The latch release had to be sufficiently high so a child could not reach it. The section on the right of this two-piece 10-foot-wide gate remains stationary and the left section is used for frequent access.

Fig. 6.87. To close the gate an old colonial system was used: a cannon ball and chain. Gravity pulls the gate closed. A simulated cannon ball was made from a pipe nipple and caps, which allowed adding weight so the exact amount of metal scrap could be added to the center cavity and adjusted for proper closure without excessive force.

Fig. 6.88. A metal latch was located that would close and locate the release high enough to meet the insurance company's access requirements. However, the mounting holes were wider than the 1.25-inch gate frame so it could not be attached as designed. The solution was to narrow the mounting and weld the latch to the gate frame. To make it easier to bring the MIG welder to the fence and because it was welded outdoors, self-shielded wire was used.

child. The simple approach was to employ a latch that required lifting to open after it was shut. This could be mounted high on the inside of the gate frame. However, the metal latches available were all wider than the tubular gate perimeter frame. The bolt holes provided were too far apart to attach to the 1-inch-wide steel frame member.

The solution was to remove most of the latch section containing the bolt holes and weld the latch to the frame member. Because this was welded in place, it was easiest to use self-shielded wire and avoid moving my large shielding gas cylinder outdoors. I also did not have to be concerned about the shielding gas being

carried away in a wind gust. For this application, an .035 self-shielding wire was placed in the MIG welder. The latch and gate frame were first ground to bare metal. Sufficient current was used to ensure good penetration and short tack welds joined the latch to the frame. Unlike when welding with MIG solid wire and gas shielding, weld appearance was not as good but as with many applications, good penetration is most critical and a small grinding wheel makes the appearance acceptable.

The automatic gate closure mechanism needed little welding and was one I first observed in historical Colonial Williamsburg. It was used on several wooden gates and

employed a chain, cannon ball, and gravity! There are kits available for purchase, but this large gate needed a heavy weight so it would close but not slam shut.

The solution was to construct a simulated cannon ball that could be adjusted for the proper weight. It was assembled using a 3-inch-diameter threaded-pipe nipple and end caps. A metal loop was tack welded to the cap top that attaches to the center of the chain connected to the gate. A metal pipe was set in the ground with concrete. Then some scrap metal pieces were added to the inside cavity until the gate closed reliably without excessive force. The device has worked for 15 years without problems!

Project: Welding a Hydroformed Frame

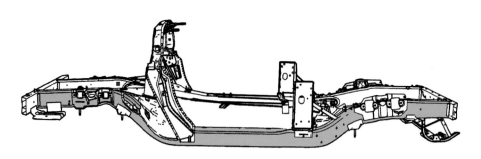

Fig. 6.89. This Corvette aluminum frame repair project reinforces some of the key points addressed about MIG welding this material. The project is repairing a hydroformed aluminum side rail used on a Z06 Corvette. (Picture adapted from **GM Collision Repair Manual**)

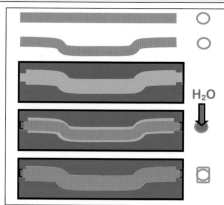

Fig. 6.90. Hydroformed Corvette frame rails start with a round tube. The tube is mechanically bent to the approximate contour. It is then placed in a die, the ends are plugged, and high-pressure fluid is pumped in to expand the tube to fit the die. The final shape is produced accurately and the material strengthened by the cold expansion. The approximate shapes at each step are shown above.

Hydroforming parts is gaining popularity in automotive applications. The Corvette has used this manufacturing method since 1997 on steel frames. When the high-performance Corvette Z06 was introduced in 2001, a similar hydroformed frame was made from aluminum. Although the new C7 Corvette frame is made from welded aluminum construction it also used hydroformed parts for the main center section.

The hydroforming process is an interesting fabrication technique that has the advantage of producing a complex shape with minimum bending and welding (see Figure 6.90):

- The process starts with a round tube.
- The tube is bent to roughly the contour of the finished frame.
- It is placed into a die and the ends plugged.
- High-pressure fluid expands the tube to fit the die.
- The plugs are removed, the fluid drained and the round ends cut off.

In addition to accurately producing the final shape, the material is strengthened by the cold forming and expansion.

Selecting the wire for aluminum welding is more complex than with steel and stainless steel. The following example shows how it can be simplified using a commonly available

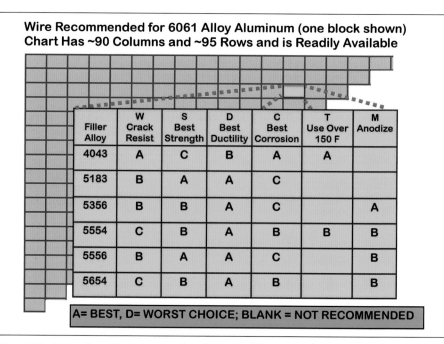

Wire Recommended for 6061 Alloy Aluminum (one block shown) Chart Has ~90 Columns and ~95 Rows and is Readily Available

Filler Alloy	W Crack Resist	S Best Strength	D Best Ductility	C Best Corrosion	T Use Over 150 F	M Anodize
4043	A	C	B	A	A	
5183	B	A	A	C		
5356	B	B	A	C		A
5554	C	B	A	B	B	B
5556	B	A	A	C		B
5654	C	B	A	B		B

A= BEST, D= WORST CHOICE; BLANK = NOT RECOMMENDED

Fig. 6.91. Selecting the proper aluminum welding wire looks complicated at first, but only one block is used for any application. In this example, the material being welded is 6061 aluminum. The row and column containing that alloy are selected from the chart (highlighted in orange here). To choose the best welding wire, determine which of the six criteria is most important.

Wire Selection for Welding Corvette Z06 Hydroformed Side Rail (Recommendation from GM Collision Repair Manual) GM Recommends Using 5356 Wire		Strength; psi	Yield; psi	% Elongation
Frame Alloy	6061-T7*	40,000	30,000	10%
MIG Weld with 5356 Wire		30,000	18,000	10%
MIG Weld with 4043 Wire		26,000	17,000	8%

*Notes: T7 refers to the material being "heat treated and stabilized." The more common T6 condition is "heat treated and aged." T7 typically has somewhat less strength but is more ductile. Always use the manufacturer's wire selection recommendations.

Fig. 6.92. Although it is not known why General Motors picked this alloy, it is considered acceptable from the chart. These are the properties of the base material and a resulting weld made with 5356. Note it is weaker than the base frame. When welding aluminum, you often cannot match the base-material strength.

aluminum welding-wire selection table (Figure 6.91). This table can be found in the *AWS Welding Handbook* as well as from manufacturers of aluminum welding wire. This table has a great deal of information in its 95 rows and 90 columns.

In the outer blocks, the base alloys to be joined are listed. The same alloys are on the top horizontal axis and the left vertical axis. This covers joining one aluminum alloy to another. If joining the same materials, just find the block that intersects the columns and row with that alloy. In the case of the Z06 Corvette, the hydroformed frame alloy is 6061.

Note that there may be other alloys used in different production years. The block having 6061 base material is highlighted and expanded and is colored orange. Within the block the filler alloy options are shown at the left, and the vertical columns define six basic criteria that can be selected, Crack Resistance, Best Strength, etc. Within the various intersecting blocks are ratings of how the specific alloy welding wires meet

the criteria. For example, alloy 4043 is rated A as having the best crack resistance but only C for strength. Note that if the block is blank the wire in that row is not recommended for that selection criteria.

The *GM Collision Repair Manual* provides a detailed procedure for replacing a portion of a hydroformed aluminum frame rail. It lists which welding wire to use for welding a new rail portion to the section that remains—wire alloy 5356 should be used.

Figure 6.91 shows that 5356 is not the most crack resistant but it has a good B rating. It does not even have the highest ductility rating but, again, has a good B rating.

In fact, 5183 has a higher strength rating. Why did GM not pick 5183? The procedure was probably tested and for whatever reason General Motors selected 5356 as the alloy of choice. You should go with that recommendation.

Figure 6.92 shows typical weld properties for 5356 wire welding 6061 base material. The resulting weld is

about 25-percent weaker than the 6061-T7 frame material. It is typical when welding aluminum to have undermatched weld and HAZ strength.

The first step in this repair is removing the composite body parts from the section of the frame needing repair. The parts bolted or welded to that frame section must also be removed, including all factory welds. (Rather than remove all of the parts from the whole damaged rail, only the damaged portion need be removed.) A vertical cut is made and the damaged section removed. Figure 6.93 shows the new rail cut in the same area and the section that will be retained.

General Motors recommends the edge of the remaining old frame section and the new piece be beveled at 60 degrees. This makes the included angle for the single bevel joint 120 degrees.

General Motors has an interesting method of making a backing plate for the beveled joint. It recommends removing a 2-inch section from the cut end of the remaining new rail that is not used. This is cut into four pieces as noted in Figure 6.95. Then 1 inch of these sections is inserted into the existing frame and carefully positioned for a tight fit. All saw blades, grinding and sanding disks should only be used for aluminum. A dedicated stainless (not steel) wire brush is used to clean all surfaces.

General Motors specifies all welding be performed with .035-inch-diameter 5356 wire with pure-argon shielding gas. Also recommended is using a pulse-arc MIG welder, probably because it is easier to make quality awkward-position welds using pulse-arc MIG. Pulse-arc MIG welding also allows lower total heat input because the average current and voltage are lower.

Because the alloy being used is heat-treated, General Motors also recommends that a two-pass weld procedure be employed, a root pass and cap pass. With both welds, after making a 2-minute weld, a 2-minute cool down period should be used before welding again.

If using sections of Corvette side rails for a street rod project and the assembly can be rotated, MIG spray arc should be satisfactory (unless you do a lot of this type of repair, then a pulse-arc welder is recommended). The heat input is higher, however, resulting in lower strength. If all welds are in the flat position, using more weld passes and a higher travel speed could offset the higher heat input.

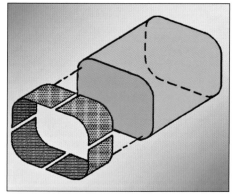

3 Fig. 6.95. General Motors recommends a unique way to make good-fitting backing bar. Cut a 2-inch section from the new rail that will be discarded. Then cut that piece into four sections as shown. Insert a 1-inch section into the old frame end and tack weld in place. General Motors recommends a gap of the material thickness (3/16 inch) be left between the frame rail pieces. (Picture adapted from **GM Collision Repair Manual**)

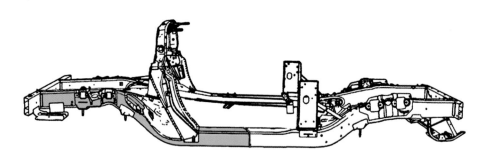

1 Fig. 6.93. To replace the section needing repair, cut and remove it from the full frame rail. A completely new frame rail must be purchased. Cut the damaged frame near a straight section. Parts attached to that section must be unbolted or if welded, welds removed. The new frame is cut to replace the damaged frame section. (Picture adapted from **GM Collision Repair Manual**)

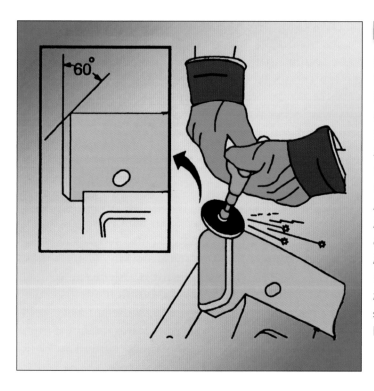

2 Fig. 6.94. Bevel both the new section to be added and the remaining old section. General Motors recommends a 60-degree angle, which leaves a large 120-degree included angle on the single-bevel joint. (Picture adapted from **GM Collision Repair Manual**)

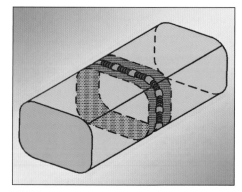

4 Fig. 6.96. General Motors recommends the use of MIG pulse arc to make the welds. Considering three of the four welds must be made out of position, this is a good choice to ensure quality. General Motors specifies 5356 alloy wire and suggests using a separate root and cover pass. Use a 2-minute wait time after 2 minutes of welding to allow the weld to cool. This reduces strength loss in the HAZ. (Picture adapted from **GM Collision Repair Manual**)

Project: Fabricating a Seat Support

1 *Fig. 6.97. A steel seat support is needed for the 1934 pro/street rod. The floor of the car is heavy, 1/4-inch-thick fiberglass, but that is not strong enough to anchor the seats or fasten the seat-belt attachment eyebolts. Bolt a steel support to the frame side rails using existing body bolts.*

Our 1934 street rod has a 1/4-inch-thick fiberglass floor that is strong but not appropriate for securing seats mounts or acting as an anchor for the seat belt anchor eye bolts. A steel floor support is needed.

The floor is flat so an H-structural arrangement can be bolted to the frame using body bolts on both sides. Figure 6.98 shows the finished brace in place.

To fabricate the structure, a T-joint is assembled from the 2-inch-wide, 3/16-inch-thick steel bar used as a crossmember. It is welded to similar-size steel bars that bolt to the frame using existing body bolts.

A quality full-penetration weld is needed between the members. A single bevel joint with a land is used to make the first weld. The plates are beveled with a grinder at a 45-degree angle making the total included angle 90 degrees.

Figure 6.100 shows the single beveled joint on the crossbar connecting to a beveled section on one side bar. Note that all surfaces to be welded are ground to clean the metal. On the side bars that attach to the frame, holes are drilled slightly larger than the body bolts. These side bars are placed outward as far as possible to accommodate weld shrinkage. The center bar is trimmed to fit tightly

between the side members. The parts are tack welded in the car; the final welds are made on a bench.

The welded assembly is turned over and a grinder, held vertically, is used to make a groove through the joint until the bottom of the first weld is penetrated. This is a common way quality welds are made to ensure the top and bottom weld passes interlock.

Figure 6.102 shows the finished assembly ready for priming and installation. The seats and seat-belt anchor eye bolts can now be properly secured as opposed to just being fastened through a 1/4-inch-thick fiberglass floor.

2 *Fig. 6.98. Bolt the seat support bar on the frame and the cross brace, which is used to fasten the seat retaining bolts and seat belt eyebolts. It is made from 2-inch-wide by 3/16-inch-thick steel.*

3 *Fig. 6.99. A full-penetration weld is desired, so prepare the cross bar and the side rail bars with a single bevel joint (shown here) using a grinder, leaving half the plate as a land. The joint is a tight butt. Use a hot single pass on what will be the top side.*

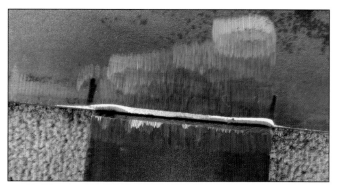

4 *Fig. 6.100. Grind the weld joint to clean metal in preparation for welding. Make no attempt to fully penetrate the material in one pass. Make a full-penetration joint with a weld on the opposite side. Bolt the side bars in place and tack weld prior to welding the assembly on a workbench. Drill the mounting holes slightly oversize, and move the side bars as far outward as the holes allow, to compensate for the shrinkage that occurs after welding.*

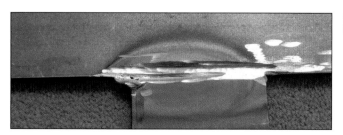

5 *Fig. 6.101. After welding both side-rail members to the cross bar, turn the assembly over to weld the back side. Hold the grinder vertically and remove material until the bottom of the first weld is reached. Make a hot weld pass from the second side that penetrates into the bottom of the first weld.*

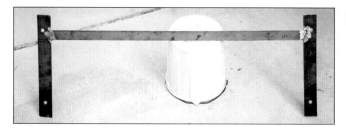

6 *Fig. 6.102. Grind the welds flush, and the finished assembly is ready to be primed. Then bolt the assembly to the frame through the fiberglass floor. It provides good support for the seats and floor-mounted seat-belt eyebolts.*

Project: Repairing a Decorative Bench

Fig. 6.103. A decorative metal bench was for sale at a local plant nursery that was perfect for an area in the yard. However it had several broken welds. It was the only one available and the owner could not get another. I was able to purchase it at half the original price. I could have paid for a small MIG welder with the savings.

Fig. 6.104. After 20 years the bench has a number of rusted areas but still serves the decorative purpose intended. Despite rust in other areas, the welds made to repair the bench are still sound. They were painted, like the bench, in matt brown to match the patina of the inevitable rust that would occur.

While looking at plants in a local nursery, my wife noticed a welded metal bench that would look good in an area of the yard. It was of complex construction with a sheet-metal mesh bottom and decorative back. It was painted brown, so when it rusted it would still look appropriate for the area intended. However, it had several broken welds and was the only one the nursery had available.

It was easy to negotiate a half-price purchase with the owner. I saved enough money to have pur-chased a small MIG welder and a small spool of self-shielding wire. It's interesting that after 20 years, parts of the bench have rusted but the welds that supported the base are still fine. It still serves the décor purpose intended.

Project: Mending a Weather Vane

Fig. 6.105. There is a pair of great blue herons on the lake where we live. We thought this metal sculpture weather vane would be fitting for the pergola on top of our dock. It was Laser cut from steel including the metal point that functioned as the swivel pivot on a metal post. Although the patina of the rusting heron was attractive the pivot point rusted and failed.

Fig. 6.106. The best way to repair the pivot was to weld a stainless steel screw to the heron body. It was welded with steel wire, which in this case produces a satisfactory weld since the small tack welds used are mostly steel with minor dilution with the chrome and nickel from the stainless screw. Even if these alloy elements were at a higher level making the weld brittle it would still work in this low-stress application.

This small welding project saved a rusted metal sculpture that sits atop our pergola. We live on a small lake and have a pergola over our dock. There is a pair of great blue herons on the lake that often can be seen sitting on top of a tree watching for fish schools near the shore. At times one sits on top of our pergola doing the same, looking for fish on the shore. We found a metal heron weather vane sculpture that would fit perfectly above the pergola. It had just a clear coating and would rust to a patina similar to the unpainted treated lumber used to make the dock. All was fine for several years until the rust wore the

pointed tip on which the weather vane pivoted. The remainder of the sculpture and weather vane function was fine.

I decided the easiest repair was to weld a thin stainless screw to replace the rusted metal pivot point. I ground off the screw head and tack welded the screw to the metal heron using steel wire. Can you weld stainless steel to steel? The answer is yes. However, it will all melt together but the remaining weld could be brittle.

In this application using small tack welds, the majority of the deposit was steel with some addition of chrome and nickel. This could lead to excess strength and a brittle

deposit if even 1/3 of the deposit were from the stainless. If that is the case, the weld could be hard and brit-tle. However, for this application and the structural strength required that is not a problem.

One industry technique used when welding stainless steel to steel is to employ a low-penetration pro-cess, such as short circuiting MIG welding to place stainless steel weld beads side by side on top of the steel plate. This is referred to as "buttering the surface of the steel." It produces a predominantly stainless layer on top of the steel. Then this "buttered" layer can be safely joined to the stain-less member.

Application: Subframe Connector

Figure 6.107 is a view of a subframe connector installed in a unibody first-generation Camaro. It ties the front and rear subframes together, stiffening the chassis below the floor. The connecter is fit into a hole cut into the floor. MIG welding is the ideal tool for making the required fillet welds to join the connector to the structural floor. The floor is welded completely around the exposed perimeter to obtain the required strength.

Fig. 6.107. This subframe connector was MIG welded to the early Camaro front and rear subframes but ties into the unibody floor. MIG welds were used to join it to the floor around the circumference.

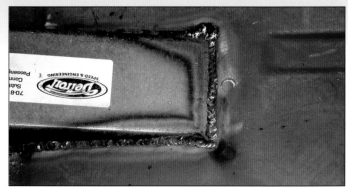

Fig. 6.108. Intermittent MIG short-arc fillet welds were made and overlapped to provide the full-perimeter weld. This approach minimizes the distortion of the floor sheet due to welding heat. The main objective is to get sufficient penetration and tie-in between the connector and the floor.

Application: NASCAR Stock Car Chassis

NASCAR specifies that a race car chassis be constructed from heavy-wall mild-steel tubing. It also defines the location of most of the tubing that make up the elaborate roll cage. NASCAR is very safety conscious and believes in a bend-before-break approach to car construction.

MIG welding is used to join most of the components. Numerous multi-tube intersections form the cage that protects the driver from roll-overs and side collisions. At 200-mph speeds, a T-bone accident could be devastating. The complex multi-tube intersection welds are made with MIG short arc. The welds are of sufficient size and are flat to slightly convex. In a technical paper

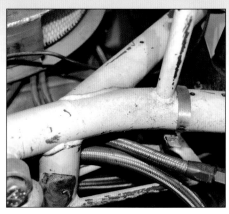

Fig. 6.109. NASCAR frames are made from mild steel tubing. Most roll-cage and frame welds are done with MIG. NASCAR advocates a bend-before-break approach to chassis and roll-cage design because the organization is very driver-safety conscious. (Photo Courtesy Bob Bitzky, ESAB)

Fig. 6.110. A 250-amp welder is of sufficient capacity and, in short arc mode, makes the required weld deposits. Here the weldor is tying in the roll cage tubing to the frame chassis member.

Fig. 6.111. NASCAR specifies tubing type, size, thickness, and location and defines the tubing arrangement for the driver safety cage. Many intersecting complex joints require welding skill to ensure good penetration.

Fig. 6.112. This complex tube intersection shows desirable weld shapes. These are the proper size and are flat, which avoids possible cracks in the center and near the weld edges. (Photo Courtesy Bob Bitzky, ESAB)

Fig. 6.113. Some of the welds, such as these on the motor-mount brackets, are made with higher-current spray arc. MIG spray arc provides greater penetration. These welds can be made in the downhand position (a maximum of about 45 degrees). (Photo Courtesy Bob Bitzky, ESAB)

published in the *AWS Welding Journal*, it was noted that ER70S-7 alloy welding wire was used in Richard Petty's shop at that time.

Figure 6.113 shows welds partially completed in a motor-mount support. All of these chassis welds are made with MIG.

An example of the quality of the NASCAR chassis is seen at the Darlington Raceway Museum in South Carolina. The museum is located on the NASCAR race track grounds. There are a number of historic race cars on display as well as an excellent hall-of-fame display of drivers and some crew chiefs.

Fig. 6.114. The Stock Car Museum at the Darlington, South Carolina, race track provides an excellent opportunity to see the results of the NASCAR bend-before-break design.

The quality of construction and weld integrity are seen in Darrell Waltrip's crashed car from the Daytona International Speedway July 1991 race. A TV telecast of the crash is shown in a video with the car display in the Daytona International Speedway Museum. The commentary is by

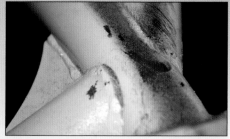

Fig. 6.115. Darrell Waltrip drove this car at the Daytona Raceway. When traveling at top speed, the car got sideways and flipped eight times as it went down the track. A video of the TV coverage is part of the car display. The car ended lying on its side, and Darrell came out with only minor injuries.

Fig. 6.116. This is a close-up view of the welds on the top of the driver-side roll cage. They were subjected to the impact as the car flipped eight times down the track. All welds held perfectly; there are no cracks. Note the small gusset used at one of the intersections.

Ned Jarrett, a former NASCAR driver, who expressed deep concern as Waltrip's car rolled over eight times down the back straightaway. Jarrett was relieved when the car landed on its side and Waltrip came out of the wreck with only minor injuries.

The welds joining the roll cage tubing contribute to the structural integrity of the assembly. None of the welds showed any indication of cracking.

A view of the right front of the car (Photo 6.110) shows that the wheel and spindle broke and separated from the car. The upper and lower A-arms are bent. However, the welds attaching steel tubing to the bottom frame rail are not broken or cracked.

Fig. 6.117. The forces on the right front of the crashed car caused the suspension parts to break and the wheel (with its spindle) to separate. The fillet welds on the chassis support tubes are still intact.

Fig. 6.118. This view of the top front of the driver-side roll cage shows excellent weld quality. Like the rear section, it was subjected to very high forces but prevented the cage from collapsing. The fillet welds have held without cracking. The weld shape is ideal, being flat to slightly convex, and of sufficient size to provide the needed strength.

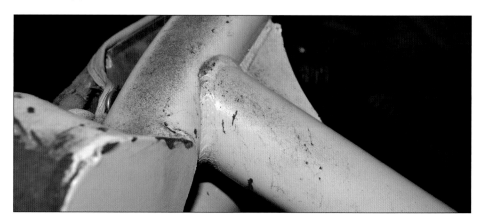

Crack Inspection

A crew chief of a Baja race car team told me about a unique method of looking for cracks in chassis welds. He mentioned that these cars are subjected to high-stress loads over a grueling 950-mile off-road race. When asked how they examine the welds after a race he explained that the frames are painted a light color similar to the Darrell Waltrip NASCAR frame. After the race they are washed thoroughly with high-pressure water, the chassis is purposely left soaking wet. After a few days, the frames and welds are examined and any cracks are readily visible as reddish brown rust spots. Not as scientific as a dye-penetrant check but an interesting and apparently effective method nonetheless.

None of the visible welds on the Waltrip-crashed NASCAR frame showed any signs of a crack. There was some rust where the paint had been scraped but no indication of cracks on these 20-year-old welds!

Since Darrell Waltrip's wreck, NASCAR has moved the driver closer to the center of the car and specified the use of additional head and neck restraints. Although there are wrecks at most NASCAR races, some occasionally similar to Darrell's, new cars have a good overall safety record. ■

Application: Street Rod Roll Bar

Fig. 6.119. A roll bar was fabricated for the 1934 pro/street rod for added safety and to support the shoulder harness. It was fabricated to be removable so it could be installed in the car after the interior was completed. The exact shape and dimensions were critical.

Fig. 6.120. S&W Race Cars supplied the .092-inch-diameter wall mild-steel tubing. It also supplied two mandrel bends so the exact height and width bar could be fabricated as the car was being assembled. A fixture was fabricated from wood for this one-time project.

This pro/street rod needed a roll bar for safety and as a shoulder-harness support. It also needs to be removable so it can be installed when the interior is completed.

This design uses a single hoop with a cross brace and one support brace to the rear. It is fabricated from heavy-wall, 1¾-inch OD, .093-inch-thick steel tubing.

S&W supplied the tubing with two 90-degree bends. This allowed all tubing to be shipped as standard freight. S&W also supplied two machined pieces of smaller tubing to fit tightly in the ID of the main bar. These short sections made perfect alignment devices and backup for the upper main hoop welds.

A wood fixture was made to hold the pieces in proper alignment. The expected shrinkage from welding the top pieces as well as the cross brace was estimated. Unfortunately not enough was allowed, but this was solved by arc straightening (see Chapter 7).

The weld joints for the top of the bar are single bevels with no nose and gapped over the machined-steel

Fig. 6.121. S&W also supplied two machined sections of tubing that exactly fit the ID of the roll-bar tubing. This provides an excellent backing for the two butt welds along the top of the bar. A single bevel joint was employed to join the tubing sections. The beveled edges were gapped 1/8 inch and the machined tubing inserted for alignment and as a weld backing.

Fig. 6.122. The welded main hoop assembly was ready for finishing. Welds were ground smooth, and when the top of the main hoop was finished, it appeared as though it were fabricated from one tube. Final sanding made it ready for priming.

Fig. 6.124. It's important to carefully measure the center support cross bar, and use a metal hole saw to make the needed fishmouth joints for the center cross bar. These joints fit with very close tolerance. MIG short arc was used to weld the top butt welds and cross bar to the main hoop.

Fig. 6.123. The finished, primed bar was test fit in this case. There is just enough clearance. The rear support bar was fabricated so the main hoop is perfectly perpendicular to the floor.

weld-backing sections supplied by S&W. Two weld passes are used: a hot root pass and a cover pass. The finished welds are ground smooth.

To fit the cross brace to the hoop, a fishmouth shape was made with a metal hole saw in a drill press. The pilot holes are carefully measured and drilled, so the saw is properly located. The intersection is beveled slightly, but large fillet welds are made so the joints can be ground smooth. (This is a street/show car, not a race car.)

The primed bar is put in place for a trial fit and the back brace tubing is fit to the main hoop so it is perpendicular to the floor.

Project: Modifying an Exercise Machine

I've been using a weight machine for more than 25 years. It is well built from a company in Canada that fabricates commercial exercise apparatus employing roller bearings and pulleys. Its leg press add-on device is well built, but the suggested biomechanics says that the toes should be about as high as the knees. This system has the feet positioned much lower. I decided to add a 14-inch-high extension to the footrest. These were made from 1/4-inch-thick plate; the same as the original.

I made a MIG fillet weld to join the pieces. They were probably strong enough as the maximum load per foot is 210 pounds. The combined load is 420 pounds maximum. For those doing leg presses on a common 35-degree angled leg press machine employing steel weight plates, that would be 715 pounds' equivalent weight. It would be the same as pushing 420 pounds straight up. This is achieved with a 2:1 pulley arrangement, doubling the load. The problem isn't the strength of the joint but a characteristic of steel regardless of its strength called elasticity. (See Figure 4.60 in Chapter 4 for a full explanation.)

To stop the excess bending, a support is needed to make the assembly stiffer. I had some 1-inch pipe that was welded to the rear of the upper plate to stop any excessive bending. See Figure 6.126. I used an interesting technique to join this round pipe to the plate.

Currently, there is a movement to use this generic approach more frequently and it has been named "additive welding manufacturing." In this case rather than make a number of small gussets that could have been welded in place to support the tube, they were made of all-weld metal. I simply made a series of tack welds on top of one another and penetrated into the plate and tube. I made five triangular-shaped welds per side that provided the gussets.

Refer to Figure 6.127 for an example: A raised flange is required on a long shaft. Traditionally you could start with a bar as large in diameter as the flange and machine everything away leaving a flange. With additive welding manufacture, MIG welding can be employed to build up a weld deposit on a smaller-diameter shaft (weld bead on top of weld bead). When the height and width are sufficient, the shaft and raised welded area are machined back to the needed shape. Much less scrap is generated. This is particularly economical when expensive metals are needed.

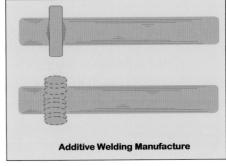

Additive Welding Manufacture

Fig. 6.125. The leg press device on my exercise weight machine had the foot locations lower than proper biomechanics suggests for optimum joint loading. A 14-inch plate was added to raise the foot placement. A MIG fillet weld was used to join the two 1/4-inch-thick plates. The strength was satisfactory but the assembly was not sufficiently stiff.

Fig. 6.126. A 1-inch-diameter pipe was welded to the back of the upper foot support to provide the needed stiffness. Although small gussets could have been used, a total of 10 were made from weld metal using overlapping tack welds. This provided the strength needed for what is equivalent

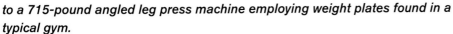
to a 715-pound angled leg press machine employing weight plates found in a typical gym.

Fig. 6.127. Using weld metal to reduce scrap and improve productivity is gaining in popularity and acceptance. It is referred to as additive welding manufacture. An example is this shaft with a flange. Rather than start with a large shaft and machine away a great deal of metal, these overlapping MIG weld beads can be deposited then the assembly subsequently machined to the finished size. This significantly reduces scrap and, especially for more expensive materials, is a viable manufacturing process.

Project: Adding a Pull-Up Device to an Exercise Machine

Although this particular weight machine has many options for different exercises, it did not include a simple pull-up bar. I decided to fabricate a neutral-grip pull-up arrangement that bolted to the rear vertical support.

I used heavy-wall 2-inch tubing and 1½-inch pipe for the handgrips. I used MIG fillet welds to join the assembly. It's probably a stronger design than needed. I drilled holes in the tubing to accept the 1½-inch heavy-wall pipe so it was welded on the outside and tack welded on the outer edge on the inside of the tube. It is very rigid and works fine.

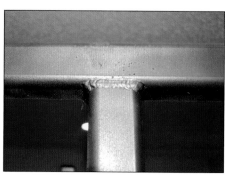

Fig. 6.128. This industrial-quality weight machine has many features but lacked a pull-up station. The one shown in this figure was fabricated from heavy-wall square tubing and pipe. It was bolted to the rear vertical support frame of the weight machine.

Fig. 6.129. MIG welds join the cross tube to the vertical upright. Solid MIG wire was used with argon-based shielding gas. Weld penetration was excellent, making for a very strong joint. The rounded edge on the horizontal bar provided a bevel to assist with the weld penetration.

Fig. 6.130. A hole was drilled into the horizontal square tubing to accept the heavy-wall pipe. A fillet weld was used to join the pipe to the front of the tubing. A short tack weld was used to secure the pipe to the inside back of the tubing.

Project: Building a Trailer Hitch

I needed a hitch for my C6 Corvette to pull a lightweight trailer. The car is equipped with what Chevy calls an NPP option, which opens butterfly valves on two of the exhaust pipes coming from the mufflers. The valves activate at above 2,800 rpm providing open exhausts and an additional 6 hp for the LS3 engine. (The sound

Fig. 6.131. A commercial hitch is not available for the C6 Corvette that is equipped with Chevy's exhaust cutout option. Apparently, the vacuum diaphragms and linkage (above the center exhaust tips) interfere with the designs.

Fig. 6.132. Fabricate the hitch from a 3/16-inch flat plate that bolts to the front vertical side of the rectangular-tube rear crossmember. Bolt a short section of the same 3/16-inch material to the bottom of the rear crossmember. Weld two pieces of 3/4-inch angle iron to the edges of the back plate to add stiffness and strength to the assembly.

is worth the price if not the extra 6 hp to achieve 436.) Unfortunately, none of the hitch manufacturers offers one to fit a Corvette with the NPP option. (I assume the vacuum diaphragm and levers that activate the valves get in the way of their designs.) In addition, the trailer was light and a heavy-duty hitch was not needed. The solution? Make a MIG-welded bracket to bolt to the rear crossmember.

I designed a simple hitch that works fine for the lightweight trailer. It is fabricated from 3/16-inch steel reinforced by welding 3/4-inch angles to the edges.

I purchased a hitch receiver and cut it to the length needed. The large metal bracket bolts to the back of the Corvette box-shaped rear crossmember. The small bracket mounts perpendicular to the rear bracket and is fillet welded. The small bracket bolts to the bottom of the rear crossmember.

After tack welding and making small fillet welds on the bottom bracket the assembly was brought to my friend and colleague Bob Bitzky, who has an enviable home shop with a 350-amp MIG welder, TIG welder, and plasma cutter.

The hitch receiver has a heavy wall thickness that needs large well-penetrated fillets. This required more amps than a small portable welder could deliver. Bob used 225 amps and made excellent fillet welds that penetrated the receiver wall.

All surfaces and the edge welds that attach the 3/4-inch angle to the 3/16-inch back plate were ground. The finished assembly has a professional appearance. It was ready for priming and paint.

The hitch installs between the center exhaust pipes and does not interfere with the muffler butterfly-

Fig. 6.133. Use a magnetic welding fixture to hold the bottom bracket and the receiver perpendicular to the rear plate. Make tack welds and attach the bottom bracket with two fillet welds, one on each side. In this case more amperage was needed than was available from my small MIG welder to obtain a quality weld on the heavy wall-thickness receiver.

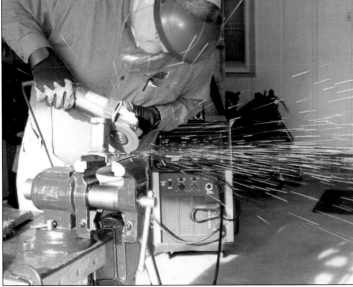

Fig. 6.134. Use a 250-amp MIG welder to join the receiver with fillet welds around the perimeter. These well-penetrated welds properly attach this critical part.

Fig. 6.135. Reinforce and grind the welds attaching the 3/4-inch angles to the sides of the rear plate. These angles provide stiffness and strength to the assembly.

Fig. 6.136. The finished assembly has the correct-size welds to match the part thicknesses. On heavy sections, such as the hitch receiver, there is no substitute for welding with higher amperage, which ensures sufficient penetration into the thicker section.

valve activation mechanism. It is in a perfect location to accept a Chevy emblem hitch cover when not in use.

Figure 6.138 shows the system with a 1¼-inch ball mount attached. The bar has a raised offset to match the height of the trailer. It works fine for a small light trailer but is not designed for vertical loads such as a bike mount or for pulling a heavy trailer.

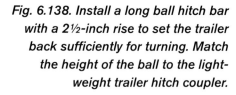

Fig. 6.137. The finished hitch assembly painted and installed. It fits neatly between the inner exhaust pipes and does not interfere with any part of the exhaust system. Therefore, the vacuum diaphragms and linkage that open the butterfly valves on the center exhaust pipes operate as designed and provide a straight exhaust path through the mufflers. They activate at about 3,000 rpm, adding 6 hp, but the sound alone is worth the option price.

Fig. 6.138. Install a long ball hitch bar with a 2½-inch rise to set the trailer back sufficiently for turning. Match the height of the ball to the lightweight trailer hitch coupler.

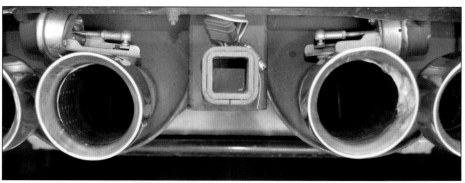

Application: Spot Welds

Fig. 6.139. Spot welds are an important MIG welding application. These are on the inner sheet metal of a unibody drift car. Drift cars require a very rigid chassis so they can respond instantaneously to driver input. Hundreds of spot welds make a single rigid assembly of the many internal body parts. (Photo Courtesy Jim Harvey, Harvey Racing Engines)

Hundreds of spot welds are used to stiffen the unibody chassis for a drift race car by Harvey Racing Engines. Drift cars need a very rigid chassis, so the driver can feel the car's movement and respond accordingly. Spot welds provide sufficient structural strength for many applications.

They can be made manually by triggering the gun for a short time. Or, if making spot welds to join overlapping sheet metal, the ideal accessory is a spot-weld nozzle that mounts on the MIG gun.

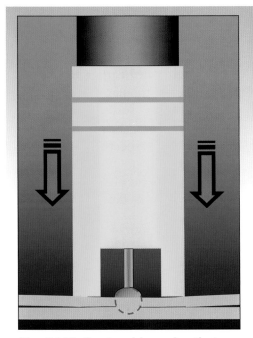

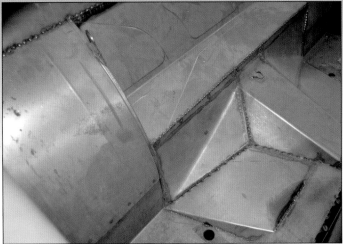

Fig. 6.141. This Chevelle street rod has a tube chassis. The interior panels of this Chevelle pro/ street rod have hundreds of spot welds joining the various hand-fabricated inner panels, which were fabricated by Legacy Motors.

Fig. 6.140. Spot-weld nozzles that fit a MIG torch can help ensure the top plate is in close contact with the bottom sheet. This illustration shows how they work. Usually made from heavy-wall copper, they have openings around the bottom to allow shielding gas and some spatter to escape. Pressure is applied to push the plates together before pulling the gun trigger.

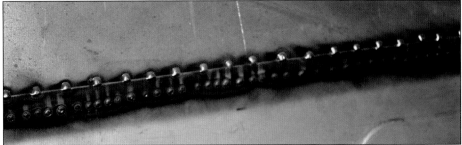

Fig. 6.142. This line of spot welds was used to join a flange joint. It provides sufficient strength for the application and will be covered with seam sealer before the interior is completed.

Fig. 6.143. The Chevelle has a rigid full-race chassis, so the sheet-metal interior panels are nonstructural. Legacy Motors fabricated the rear end by using racing gears and 36-spline axles.

Fig. 6.144. This is the back side of a spot-weld made on the opposite side to attach panel stiffeners. The low-heat input of these spot welds leaves the sheet metal free from the distortion that would have occurred if even low-heat input MIG short arc were used to made continuous welds.

A MIG spot-weld nozzle is longer than a normal MIG gun nozzle. It has openings at the bottom to allow spatter and shielding gas to exit. It is usually made of thicker material and can be pushed down on the upper sheet so it contacts the lower sheet before pulling the gun trigger. A spot-weld timer built into some welders can help ensure the proper amount of weld penetration without burning through.

Figures 6.141, 6.142, and 6.144 show spot welds used to join Chevelle floor and inner body panels. The car has a tube chassis and the sheet-metal interior will be covered. The spot welds provide sufficient structural strength and avoid distortion, which would occur if full welds were attempted to join these sections.

Troy Spackman, one of the owners of Legacy Motors, is building this race/show car. He will use seam sealer to cover the spot welds prior to interior finishing.

As mentioned earlier, this Chevelle has a full tubular race car frame with coil-over suspension and a hand-fabricated, heavy-duty axle assembly. The sheet-metal interior was fabricated to fit the custom frame. There is no need for any additional structural support from the interior panels.

Application: Triumph TR3 Floorplan Installation

Bruce Walters purchased a 220-volt input, 175-amp Lincoln MIG welder when he started a restoration project on a Triumph TR3. He had not welded since attending Purdue University more than 40 years before. However, he quickly learned the skills needed to do body and frame repair.

Bruce raced a Triumph while a student and this restoration was his first venture rebuilding an old car. He built a 387- out of a maximum 400-point restoration. The car has been in many car shows around the country. It turned out so well it stays covered on the top of his garage lift until show time. To maintain the high point score, the car is trailered to shows.

After finishing the TR3 Bruce tackled the restoration of what is now a cherry TR6 with hopped-up engine, transmission, and suspension parts of the early 1970s. Some

Fig. 6.145. Bruce Walters has restored two cherry Triumphs and is now working on another as a driver. With this 220-volt input Lincoln welder, he has been able to repair sheet-metal and chassis parts.

Fig. 6.146. Bruce's first restoration plan was to bring a TR3 back to original condition. It has earned 387 out of 400 points in car shows and is trailered to maintain that high score. He added this storage lift for more room. The red TR6 was restored with hot rod parts of the day. It also is so well done that it is not a daily driver.

of the performance modifications are similar to those used when racing in SCCA. (He recalls racing at some of the same SCCA events, although in a different class, as Rodger Penske, current *Autoweek* writer Denise McCluggage and Indy racer Janet Guthrie.)

Bruce's latest TR3 project utilizes a modern four-cylinder engine and drivetrain. The car has good fenders, needing only minor straightening in addition to some minor chassis repair. Unfortunately, the floor had sufficient rust that it needed to be replaced.

Before removing the floor, Bruce installed the fabricated body support braces he built for his first TR3 project. They had been lying outside for several years and, although rusted, they fit fine. These support braces are necessary when removing the floor so the body dimensions do not change. You want the floor correctly welded to the chassis, and the floor provides significant structural support.

The replacement floor panels came coated with a weld-through primer. Since mostly small tack and spot welds will be used the weld through primer is satisfactory. However, when making structural welds on frames or other parts, the primer should be removed.

Although a number of coatings claim to be weld-through they often create weld porosity even if it is not visible on the surface. This is especially important if welding high-strength steels because any hydrogen compound that might be in the finished primer could cause cracking, not just porosity issues.

The replacement floor panel has flanged edges that overlap the driveshaft tunnel. Bruce has tried a number of methods to eliminate burn-through on thin sheet metal.

Fig. 6.147. Here is Bruce with his latest project: restoring another TR3, but this will be his hot rod. When the sheet-metal restoration is completed, the car will be equipped with a modern four-cylinder engine and drivetrain.

He found this copper backing plate with magnets in each corner works well. The copper absorbs heat, reducing the chance of burn-through. Also, weld metal that does penetrate is quickly solidified but does not stick to the copper surface.

While installing the new floor a rusted area on the driveshaft tunnel required fabrication of a small replacement panel. Bruce had tried using Cleco fasteners in the past to

Fig. 6.148. Before removing the floor, this fabricated body support brace was installed. When removing the floor, this brace provides significant structural support and helps retain the body dimensions. Bruce fabricated the support brace for the first TR3 project and, although now rusted, it worked fine.

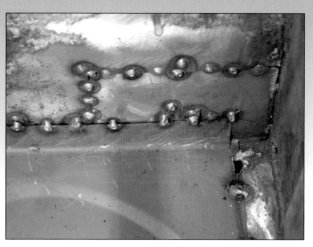

Fig. 6.150. The spot welds were used to join the new floor flange to the driveshaft tunnel. Part of the tunnel had a rusted area that was cut out, and a rectangular panel was carefully fitted to fill the removed section. Using the copper plate behind the butt seam allowed spot and tack welds to join the parts.

Fig. 6.149. This TR3 needed a new floor, which was spot welded to the driveshaft tunnel that was (for the most part) in good condition. He found Cleco clamps left too large a gap when making butt welds, causing burn-through. So he used the copper plate with magnets in each corner. He fit the replacement panel tightly and placed the copper backing behind the joint. Spot and tack welds can then be made without burning through.

hold sheet-metal panels in place in preparation for welding. However, he found that the gap, even though it was small, caused some burn-through. These devices also interfere with the use of the magnetically held copper backing plate he found to work best.

Bruce cut out the rusted driveshaft tunnel section and carefully trimmed a piece of sheet metal to fit the hole. The panel was trimmed to provide as tight a butt joint as possible. He used MIG tack welds spaced to avoid overheating and distort-

ing the sheet metal. The floor will be sand blasted, the seams sealed, primed, and painted.

The transmission cover is a separate metal section of the front floor assembly. It must remain removable to gain access for clutch replacements and adjustments. That portion of the floor was in fair condition but some areas were very thin so a replacement copy will be used. Since it may have to be removed occasionally, the full thickness of sheet metal prevents the thin sections from bending.

Fig. 6.151. The front section of the floor used a separate sheet-metal transmission cover that was only bolted to the floor. It had to be removable to gain access to the transmission and clutch when repairs and adjustments are needed.

Fig. 6.152. The front transmission cover was in fair condition and could be repaired. However, rather than spend the time, a reproduction of the original was installed. Bruce also felt there were some thin sections in the transmission cover that could easily bend if removed too often.

Application: Street Rod and Race Car Tubing

S&W Race Cars stocks a wide variety of alloys and shapes of tubing for street rod and race car fabrication. S&W supplied the material for our pro/street rod roll bar and bent the tubing to our specifications. In addition to supplying tubing and tubing kits for chassis and roll cages, the company also fabricates finished chassis and parts to customer specifications.

Figure 6.154 is a short arc MIG weld made by moving the gun forward, then pausing. This technique makes the final weld look more like a TIG weld. When TIG welding, pausing is natural when the metal rod is dipped into the puddle. Some weldors prefer this TIG-like appearance.

A similar weld appearance may be seen on an inexpensive mass-produced aluminum bicycle frame. A technical paper by an Asian welding company showed how they developed a technique using low-current

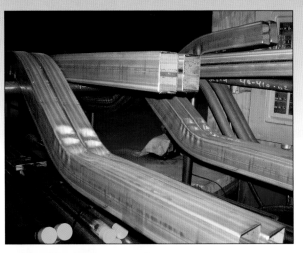

Fig. 6.153. S&W Race Cars supplies various round and square tubing for race car and street rod construction. It also supplies bent parts and can bend and fabricate to custom specification. MIG welding is the most common way to join the carbon steel parts (Photo Courtesy S&W Race Cars)

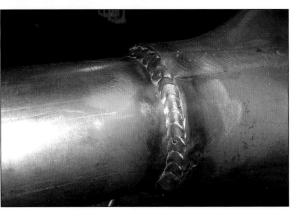

Fig. 6.154. This MIG weld is joining two sizes of tubing, one inserted into the other. The technique used is to make overlapping short welds. This avoids overheating the weld area, and for racing this works well. Note the very good wetting at the weld edges. (Photo Courtesy S&W Race Cars)

Fig. 6.155. A technique of moving the torch forward and pausing (similar to TIG welding) is used to make large fillet welds on these heavy suspension parts. Some weldors prefer to use this technique because it produces a uniform, good-looking weld, but it may also produce some additional spatter as seen here. (Photo Courtesy S&W Race Cars)

Fig. 6.156. A MIG gun manipulation technique of advancing forward and pausing created the overlapping weld appearance. The smooth welds on the top and sides were made with MIG spray arc. (Photo Courtesy S&W Race Cars)

MIG pulse arc on a robot welder to purposely match a TIG stack-of-dimes weld appearance.

MIG fillet welds are ideal for joining relatively heavy frame structures. They can quickly be made sufficiently large to provide the required strength.

Several types of MIG welds are visible on the frame component in Figure 6.156. MIG spray-arc welds have a smooth surface. These hot weld deposits are made by rotating the assembly so the weld deposits are made in the downhand position. If this higher current deposition mode can be used, there is no need to manipulate the MIG gun. Just keep the gun almost perpendicular and keep travel speed steady and smooth. Other welds were made using MIG short arc, some with the pause, stepping approach.

Fig. 6.157. TCI supplies many finished frames for 1930s through 1950s cars. At one time, the company only used TIG welding. However, TCI currently uses quality MIG welding to join many components. (Photo Courtesy Ben Bryce, Total Cost Involved)

Application: TCI MIG Welds

TCI in California produces hundreds of chassis parts and complete chassis for 1930s Fords, Mustangs, Camaros, Chevelles, and other mid-1950s car suspension parts. When the chassis on my pro/street rod was fabricated by TCI in 2000, all TIG welding was employed. In addition to TIG welding, TCI now also uses high-quality MIG welding. The front suspension coil-spring shock tower supports are joined to the frame with MIG welds.

TCI uses a somewhat unique rear crossmember that combines brackets for four-bar link suspension with a driveshaft retention loop in case of a rear U-joint failure. This is a natural for high-quality MIG fillet weld joints that can be made much faster

Fig. 6.158. Here is a good example of MIG welding providing quality fillet welds on heavy section frame components. TCI uses a unique rear crossmember that supports the four-bar link brackets, and these have been drilled for a wide range of adjustment. The center opening doubles as a rear-driveshaft retention loop. (Photo Courtesy Ben Bryce, Total Cost Involved)

and with similar weld quality to TIG for this steel application.

Making large fillet welds is much easier and quicker with MIG welding. Filler-metal deposition rates are many times higher and MIG short arc weld metal is easy to hold in position.

In Figure 6.160 MIG fillet welds are shown joining intersecting brackets, an excellent process choice. MIG also allows the proper-size fillet weld to quickly attach the suspension bracket to the side frame rail. All welds were made manually and high skill can be seen in the uniformity of the weld deposits.

Heavy, thick material is required for constructing the four-bar link bracket that attaches to the rear crossmember. It has many drilled holes for adjusting suspension geometry. The bracket is joined to the frame

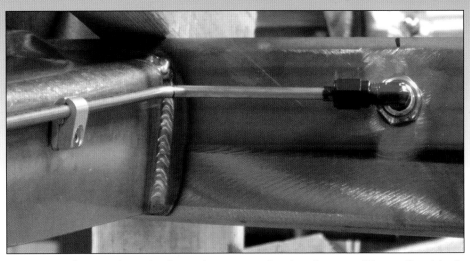

Fig. 6.159. Connecting a horizontal crossmember to a frame rail is another ideal place for a MIG weld. TCI supplies many options, such as the fully plumbed brake lines shown here. (Photo Courtesy Ben Bryce, Total Cost Involved)

with properly sized quality MIG fillet welds.

An excellent example of the best use of MIG and TIG welding processes is presented in Figure 6.162.

The corner joint edges of this coil spring/shock tower were welded with TIG while the center hat section is joined with a well-made MIG fillet weld.

Fig. 6.160. MIG welds are also used for joining suspension brackets to the frame rail. The quality of the workmanship is obvious. TCI supplies a number of different options, from classic solid axles to coil-over systems with adjustable shocks. (Photo Courtesy Ben Bryce, Total Cost Involved)

Fig. 6.161. A close-up of this fillet-welded bracket shows the quality and uniformity of welds on this four-bar link suspension component. TCI has an extensive machining department, so fit-up can be controlled, which is key to making quality welds (Photo Courtesy Ben Bryce, Total Cost Involved)

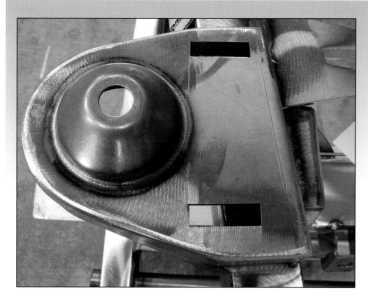

Fig. 6.162. A combination of MIG and TIG welding was used for this component. This spring shock tower shows the uniform MIG weld attaching the hat section to the top assembly. TIG welds join the outer edges of the assembly. The suspension part is joined to the frame side rail with a MIG fillet weld. (Photo Courtesy Ben Bryce, Total Cost Involved)

Application: Rear Quarter Panel Replacement

MIG welding is seen joining a replacement rear quarter panel to the original fender in Figure 6.163. A rusted section of the quarter panel was cut away. A matching new panel section was trimmed to fit the opening. Overlapping spot welds join the new panel section to the rust-free area of the original fender.

Each spot weld is allowed to cool before coming back to the same area to place another weld over the previous one. This approach minimizes distortion of the sheet metal.

The raised spot welds on this repaired quarter panel have been ground to almost the base metal. A few of the overlapping spot welds are still visible in Figure 6.164. A light coat of body filler followed by block sanding lowers any high spots and fills minor low areas. That provides the desired flat finish without making any area excessively thin.

Fig. 6.163. A section of a rusted quarter panel was cut away and a replacement panel was spot-welded in place using MIG. Overlapping spot welds were spaced along the seam, and once the area was cooled, additional spot welds were placed overlapping the previous welds. (Photo Courtesy Year One)

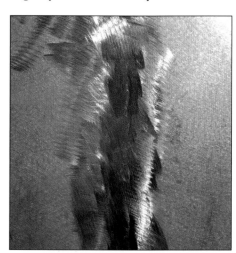

Fig. 6.164. Spot welds were ground smooth until almost flush with the base metal. A light coat of body filler filled minor low spots and block sanding lowered any high areas to provide the desired smooth surface.

ADVANCED MATERIALS AND
METALLURGICAL PROCESSES

By employing advanced high-strength steels, automakers are able to significantly improve crash test rating performance. These new steels are used for side beams, rocker panels, and B-pillars for strengthening the chassis. They are also used to reduce weight in floors, center consoles, and components such as seat backs. The small addition of alloy, such as carbon, significantly increases steel strength. The testing of metallurgical properties and investigating weld defects provide valuable information about the welding process.

High-Strength Steels

A number of the newer high-strength automotive steels use very small amounts of alloys that would cause the steel to be very brittle if used in quantities exceeding very low levels. For example, some of these steels use boron as an alloy in quantities of only .005 percent. Other microalloys are titanium, vanadium, niobium, and nitrogen that are often used in amounts less than .01 percent.

Many micro-alloyed steels are not recommended for welding because the HAZ can be significantly weakened, and/or cracks may form, and therefore the part may structurally fail. In general, car manufacturers

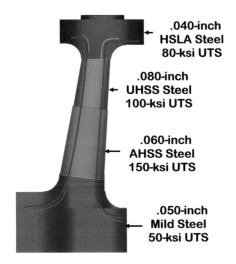

Fig. 7.1.This example shows how new high-strength and ultra-high-strength steels are used to fabricate a composite door B-pillar from a tailor-welded-blank. Employing advanced high-strength steels, automakers significantly improved crash test ratings. Many of these micro alloyed steels are not recommended as weldable since the HAZ significantly weakens them or cracks may form.

.040-inch HSLA Steel 80-ksi UTS

.080-inch UHSS Steel 100-ksi UTS

.060-inch AHSS Steel 150-ksi UTS

.050-inch Mild Steel 50-ksi UTS

Steel Type	Tensile Strength (ksi)	Yield Strength (ksi)	% Elonga-tion	Key Chemical Elements
Mild Steel	50	35	27%	C, Mn, Si
HSLA (High Strength Low Alloy)	80	65	24%	Low C, N, Va, N, Cu
UHSS (Ultra High Strength steel)	100	80	18%	Boron
AHSS (Advanced High Strength Steel) DP (Dual Phase) 900 GN	150	95	10%	C, Mn, N, Ti, Al, Va, Ti, Cu
Mart 950 (Martensitic Q&T)	170	135	6%	>.20 C, Mn, Si, Cr, Mo, B, Va, Ni

Fig. 7.2. Some DP steels are bake-hardenable, meaning strengthening occurs after the steel goes through a paint-bake cycle. Between bake hardenability and the higher level of strength achieved when cold formed, DP steels can increase yield strength about 20 ksi. Some of the highest-strength steels use boron as an alloy, resulting in a tensile strength of as much as 230 ksi.

recommend arc welding for some of these steels, but the heat input must be kept to a minimum.

Figure 7.1 shows a door pillar fabricated from a tailor-welded blank made from four different-strength steels. Tailor-welded blanks are often fabricated from several different material types and thicknesses. They may also have different coatings, such as door bottoms with aluminized or other coatings to help resist corrosion in rust-prone areas.

These different materials are often joined with automatic welding using lasers. Laser welding uses very low heat input and creates a minimum-size HAZ. They are often welded into large sheets that can be put into stamping dies or cut to the shape and size as needed.

The fact that weld quality is acceptable with laser welds does not mean arc welding, with its higher heat input, is acceptable. Note: Some high-strength steels require heating to high temperatures to form the contours used to stiffen the part. The car manufacturer must define the weldability of steel types because each can be manufactured in a number of different ways, depending on the steel supplier.

For example, manufacturers may indicate the part may be welded, for collision repair, only in mild steel or HSLA (high-strength low-alloy) steel sections. Even then, they may specify a number of small spot welds be used. These steels have been used to fabricate chassis parts for some trucks, and the labels on these steels state that they are made from high-strength steel and welds should not be placed on the material.

Some dual-phase (DP) steels are bake-hardenable, meaning strengthening occurs after the steel goes through a paint-bake cycle. HSLA steels do not exhibit this characteristic and are usually considered more readily weldable. For DP steels, it is common to see an increase in yield strength of about 20 ksi after forming and baking. In comparison, HSLA steels may have an increase of about 3 ksi due to a similar amount of work hardening.

Enhanced energy absorption during a crash is another DP steel feature. For a specific yield strength, the DP-steel tensile strength is higher than that of HSLA steels, which enhances crash performance. For equivalent crash performance, a DP steel may allow reducing material thickness over an HSLA steel by about 10 percent.

Work-hardening response to deformation is different for HSLA and DP steels. HSLA steels begin to lose formability as soon as deformation starts. DP steels maintain formability into the press stroke and better distribute the strains across the part.

Some of the highest-strength steels use a small amount of boron as an alloy, which produces very high strength. This ultra-high strength is useful for a number of components, which must be hot formed at temperatures exceeding 1,700 degrees F and quenched to achieve their strength. The resulting tensile strength of the component is as much as 230 ksi.

Applications for the ultra-high-strength steel include B-pillars, center tunnels, inner sills, and firewall beams.

Welding many of these steels is generally not recommended. If you feed silicon bronze MIG wire in a similar fashion as with steel wire, the alloy melts at about 1,600 degrees F, well below that of steel. Your braze joint is not as strong as a fusion weld but may be satisfactory for some applications. This avoids melting the base material, which may avoid weld cracking but still reduces the strength of the base material in the HAZ.

Alloys in Steel

Steel is made of iron (generally more than 97 percent) and a variety of other alloying elements. Relatively pure iron has a tensile strength of less than 20 ksi. Small additions of alloys significantly increase this strength. The following describes two ways that alloys affect properties: solid solution strengthening and unintentional additions.

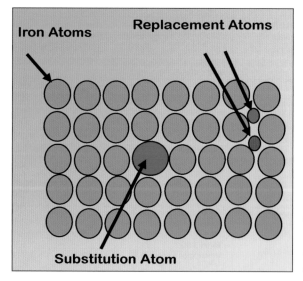

Iron Atoms
Replacement Atoms
Substitution Atom

Fig. 7.3. Alloy elements strengthen iron to make steel. A carbon atom is smaller than an iron atom, and strengthens iron by sitting between the iron atoms. This increases the force required for rows of iron atoms to slide over one another. Other alloying elements, such as copper, phosphorous, and vanadium, have larger atoms that replace the iron atom and strengthen by substitution.

Solid Solution Strengthening

A carbon atom is smaller than an iron atom and strengthens the iron by sitting between the iron atoms. As a result, the rows of iron atoms require more force to slide over one another. This sliding or snapping from one set of atom binding forces to the next in a crystal is why materials yield and ultimately break. Only a small amount of carbon is needed to significantly increase this force. Most weldable steels have about .10- to .15-percent carbon. The 4130 chrome-moly steel has .30-percent carbon and 4340 has .40-percent carbon, which is considered high as related to weldability.

Copper, phosphorous, and vanadium are other alloying elements that have larger atoms to replace the iron atom for higher strengthen. Adding alloying elements is known as solid solution strengthening. A combination of alloying elements and proper heat treatment can increase the strength of simple iron from less than 20 to 250 ksi (for moderately alloyed 4130, for example).

Unintentional Additions

Hydrogen is not an intentional alloying element. In fact, it causes problems when welding steel and can ultimately lead to weld cracking. Some elements, such as hydrogen, sulfur, and arsenic, are not wanted but come with the base plate, welding wire, or welding procedure.

As an example, hydrogen can come from a number of sources: moisture in the air that is pulled into the shielding gas stream; hydrocarbon-drawing lubricants on the welding rod or wire; and paint, oil, and grease contamination on the base material.

Hydrogen is readily absorbed in molten weld metal, but it is rejected from solid steel. Once the weld solidifies, the very small hydrogen atom diffuses quickly through the cooling weld into the surrounding base metal. In highly stressed areas, such as nonmetallic inclusions or martensite that can form in higher-carbon steels like 4130, hydrogen can accumulate. If the concentration becomes sufficient, it causes a crack to form.

When the weld cools below about 350 degrees F, hydrogen diffusion occurs, but at a much slower rate. Hydrogen cracking or delayed cracking, most often starts in the base metal because it takes time for the hydrogen to create the high-stress levels to cause cracking. Keep in mind that the higher the material strength, the more hydrogen can be a problem. Since a weld shrinks along the joint length as it cools, the stresses are pulling along that axis. Therefore, hydrogen delayed cracks are often transverse to the weld, and hence they may start in the base material and go completely across the weld joint.

Hydrogen can be observed evolving from a cold weld using a method that was once employed to measure hydrogen levels created by welding materials. Here's an experiment that will allow you to observe this occurring:

1. Purchase about a quart of glycerin. Put it in a large clear glass or small glass bowl of a size that holds a small steel test plate. A 2 x 3 x 3/16-inch-thick piece of steel can be used.
2. Make a 2-inch-long stick weld on the steel test plate (a MIG weld can be used but has less diffusible hydrogen).
3. Quickly remove the flux, pick the test plate up with pliers, and quench it in water. That significantly slows the rate of hydrogen diffusion.
4. Dry off the test plate and put it in the glass or bowl of glycerin.

The hydrogen diffuses slowly out of the weld and forms bubbles on the surface. The bubbles increase in number and size for several days. They rise slowly in the thick glycerin before escaping into the atmosphere. The amount of glycerin used should cover the sample by several inches for the best observation.

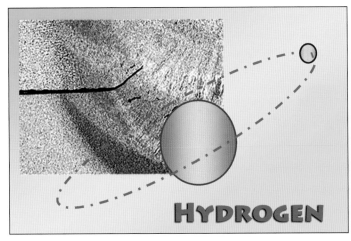

Fig. 7.4. Hydrogen, like other elements such as sulfur and arsenic, are not wanted but are introduced with the base plate, welding wire, or welding procedures. Hydrogen is readily absorbed in molten weld metal; however, it is rejected from solid steel and causes delayed cracking. The higher the material strength, the more hydrogen can be a problem. Hydrogen-delayed cracks are often transverse to the weld.

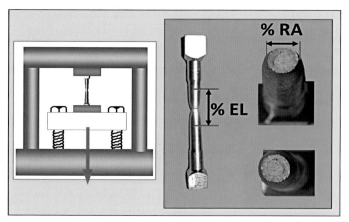

Fig. 7.5. Two important material properties are ultimate strength and yield strength. A round bar of the material to be tested is pulled in a large press and the load that is required to break the bar is measured (on the left). Percent reduction in area (RA) and percent elongation (EL) are measured from a broken tensile test specimen and relate to the material's ductility.

Use a low-hydrogen welding process to avoid hydrogen cracking, especially when welding high-strength steels, such as 4130 chrome-moly. TIG welding provides the lowest hydrogen in welding, but the TIG rod should be free of oil and drawing lubricant, so higher levels of hydrogen aren't introduced into the weld.

Use the paper wipe 11test mentioned in Chapter 5. An advantage of TIG welding is that the arc radiation burns off some of the rod-surface contaminants before the melted drop goes into the weld pool.

MIG welding is the next lowest process relative to its ability to produce low-hydrogen deposits. However, the MIG wire leaves the gun's contact tip at a fairly high rate of speed, which does not allow time for surface contaminants to burn off the wire before it enters the weld. Stick welding produces the highest amount of hydrogen, especially the E60XX- series electrodes. E7018 is designated as low hydrogen and produces far less weld hydrogen than an E6010.

When welding 4130 chrome-moly, remove all mill scale as it also contains moisture. When welding high-strength steels, the surfaces to be welded should be ground to clean metal and a solvent used to be sure all oil is removed.

Metallurgical Property Tests

A number of tests are often performed to determine weld quality. Some of these tests are specified as required for qualifying a weldor's skill or welding procedure (i.e. amperage, voltage, travel speed, joint design) for a specific application.

Tensile Test

Two important material properties are ultimate strength and yield strength. Ultimate strength is the maximum load a test specimen can withstand before breaking. Figure 7.5 shows a typical tensile-test specimen. A nominal .50-inch-diameter round bar is machined from the material to be tested and pulled in a large press. The load required to break the bar is measured in pounds. The result is provided in pounds and divided by the area of the bar, and therefore ultimate tensile strength is expressed in psi or ksi.

The two broken test pieces are put together, and the amount the bar stretched is measured. Prior to testing, two marks are placed on the test bar at a specified distance. The amount the pieces moved after testing is reported as a percent of elongation. This measurement determines how ductile the material is when it is loaded slowly.

The fourth parameter is the percent of reduction in area. The diameter of the bar that tapered down to the smallest diameter is measured, and the area compared to the original area of the test bar. Smaller standard specimens are also used for thinner material, as are flat specimens for sheet-metal testing.

Yield strength is determined from a tensile specimen test, but for the purpose of this discussion, it can be considered the elastic limit. When the maximum load has been removed, the material is allowed to return to the original length before any load was applied—this is yield strength. The actual determination of the value requires measuring and plotting the amount the bar stretches as it is being loaded. Most designs require that a material never reaches the yield point when loaded. But head bolts on today's modern engines are an exception.

All four material properties (ultimate strength, yield strength, percent elongation, and percent reduction) in an area are often specified as the minimum amount a base material or weld must achieve to meet a certain specification.

Charpy Test

Toughness provides resistance to brittle failure when a material is quickly or impact loaded. The Charpy test measures this characteristic.

For the test, a bar (a cross section about 10 x 10 mm and slightly more than 2 inches long) is machined from

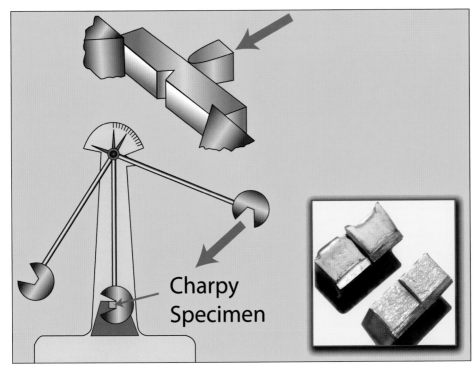

Fig. 7.6. Another very important material characteristic is called toughness, which is the resistance to brittle fracture. It can be measured with a Charpy test. A weighted hammer is used to break a Charpy test specimen and the energy required is measured. The Charpy test can also be used on thinner materials; for instance, a 1/4-inch specimen needs only a .100-inch thickness.

the material under evaluation (the specimen). A specific-radius V-notch is machined on one surface, which forms a stress concentration.

The test device contains a weighted hammer and a pendulum mechanism that pivots on a low-friction bearing. A mechanical pointer measures the height of the hammer rise. To begin calculating toughness, first find the initial potential energy of our test device (without a specimen in place):

Initial Potential Energy (ft-lbs) = Hammer Weight x Distance Raised

50 pounds x 2 feet = 100 ft-lbs

The specimen is placed on the test fixture as shown in Figure 7.6. The heavy hammer-like device is released and strikes the back of the specimen. A crack quickly forms at the root of the V-notch and breaks the specimen.

Continuing our example, the hammer rises .9 feet and 45 ft-lbs of energy were left after it broke the specimen. According to the measurements, the hammer started with 100 ft-lbs, and when finished breaking the specimen, it had 45 ft-lbs of energy left. Therefore, the specimen required 55 ft-lbs of energy to break (100 – 45 = 55).

The test specimen can be easily cooled in an alcohol bath to the required test temperature. Since steel becomes more brittle (less tough) as the temperature decreases, a material requirement is specified, for example, 20 ft-lbs minimum at 0 degrees F.

So if building a bridge in Alaska, the Charpy test may be required to be made at -40 degrees F. It is difficult to achieve high toughness at these low temperatures, but alloyed steels and matching welding wires can accomplish the goal.

The Charpy test can also be used on thinner materials. The specimen used is called a sub-size Charpy. The smallest can be 2.5 x 10 mm.

Hardness Test

Material hardness and strength are related. One way to measure hardness is by pressing a small hard ball or specially shaped small diamond into the material being tested with a fixed load and measuring the depth or diameter of the impression. To measure hardness using this method, a device (often in a lab) provides a specified load and contains a mechanism to measure the depth the impression created. Brinell, Rockwell, and Vickers are several hardness scales that are commonly used.

A Telebrineller is a simple device that measures hardness in a shop environment (see Figure 7.8). A bar of known hardness is placed in a holder over the indent ball. A hammer is used to strike the holder and

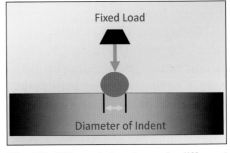

Fig. 7.7. Although hardness is different from strength, they are related. One way to measure hardness is by pressing a small, very hard ball or special-shaped diamond into the material with a fixed load and then measuring the depth of the impression.

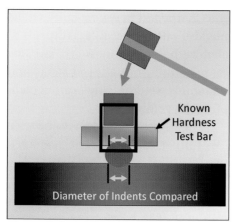

Fig. 7.8. This illustration shows a simple device used to measure hardness in the field. A bar of known hardness is placed in a holder over a very hard indent ball. A hammer is used to strike the holder and the force is transmitted through the test bar to the indent ball, then to the part being tested. The difference in indentation diameter measurements between the bar and the material being tested can be converted to a hardness value.

Brinell Hardness Standard Ball	Rockwell Hardness C Scale	Approximate Tensile Strength
135	-	75 ksi
212	-	100 ksi
315	34	150 ksi
400	43	200 ksi
-	50	250 ksi

Fig. 7.9. For steel, there is an approximate relationship between hardness and strength. Hardness, however, cannot provide an indicator of the "toughness" of the material. Some relatively low-strength steels can be very brittle.

the force is transmitted through the test bar to a very hard indent ball, then to the part being tested. A 7X magnification pocket microscope with size markings is used to measure the indent width in the bar and in the part being tested. The difference in measurements can be converted to a specific hardness.

For comparing to a specified hardness, a bar of that standard hardness is purchased, and the indent in the part is easily compared with that in the standard bar. If the indent diameter in the material being tested is smaller than the test bar, it is harder; and if larger, it is softer.

Tables are provided with a Telebrineller to allow a specific number to be defined using the ratio of the two measurements.

The Rockwell Superficial Hardness Scale is used for testing thin material. Welds and HAZs are often tested by cutting out a weld sample, mounting, polishing, and etching a cross section of it for examination in a microscope that contains a hardness tester.

Vickers values, also called diamond pyramid hardness, are often specified as maximum allowable levels for a weld and HAZ. A hardness traverse is made across the weld, and if hard, brittle constituents form in the weld or HAZ, they can be identified. These hard areas are also generally brittle and can be a cause for starting or extending cracks. Maximum hardness values are specified as acceptable levels for specific applications.

Steel has a predictable, approximate relationship between hardness and strength. Tables can be used to estimate the strength from a simple hardness reading. While this is not a perfect correlation, it's useful for a quick check. Figure 7.9 provides a correlation of approximate tensile strength with two hardness scales. Remember, however, that hardness does not provide an indicator of the toughness of the material.

Bend Tests

A guided bend test is commonly used to verify filler metal acceptability for a specific application, for an evaluation of the welding joint design and procedure, and as one requirement of weldor qualification. Figure 7.10 shows a guided bend test specimen machined to the sample size required. The weld is in the center of the specimen and is subjected to the maximum stress in the bend test.

Fig. 7.10. A bend test is commonly required when certifying welding filler metals, to test a weldor, and to evaluate welding procedures such as usable amps, volts, and travel speed. This is a guided-bend test specimen, machined to the sample size required and then bent in a fixture. The weld was made in the short direction and placed in the center of the bend.

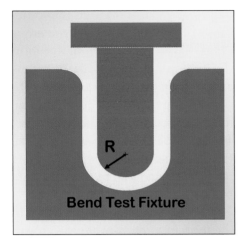

Fig. 7.11. The radius (R) is specified by welding codes and is dependent on the material thickness. It is set to provide a specific amount of elongation on the outer surface. In general, to pass the test, no cracks larger than 1/8 inch should appear.

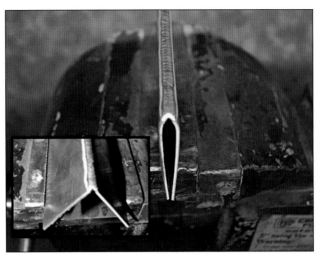

Fig. 7.12. For thin plate, such as sheet metal, a simple free-bend test is specified. This is a corner weld made in aluminum that was put in a vice to flatten the plates. This quality weld passed. The maximum crack allowed is usually 1/16-inch long for sheet-metal welds.

Figure 7.11 is an illustration of a test fixture used to make guided bend tests. The radius (R) is specified by welding codes and is dependent on the material thickness. It is set to provide a specific amount of elongation on the outer surface. In general, to pass the test, no cracks larger than 1/8 inch should appear.

To become a certified weldor for many applications, a candidate must pass a bend test in welds made in the downhand, vertical, and overhead positions, but other tests are also required.

A free-bend test is specified for thin plate, such as sheet metal. Figure 7.12 shows a corner weld made in aluminum that was put in a vice to flatten the plates. This quality weld passed.

The maximum crack allowed is usually 1/16 inch long for sheet-metal welds. Measuring the weld width before and after bending determines the amount of elongation achieved.

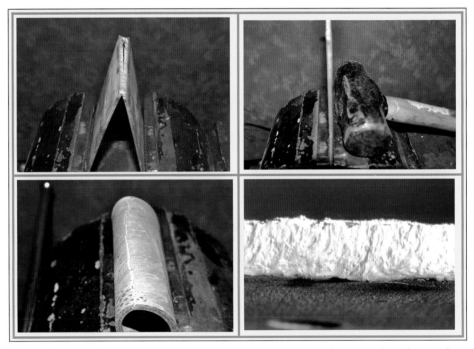

Fig. 7.13. A corner weld made in 1/8-inch-thick magnesium was bend tested. The small, unwelded end of the joint caused a crack to propagate rapidly through half of the weld. It only bent about 60 degrees before the crack formed. When the cracked weld surface was examined (lower right), it was found to be sound with no porosity evident. The fracture surface was ductile. A test of the unwelded plate was also made (lower left).

Figure 7.13 shows a corner weld made in 1/8-inch-thick magnesium being tested the same way as the aluminum weld. About 1/2 inch was left unwelded; it was understood that would create a stress concentration when the weld was tested. The first observation was that it took much more force in the vise to bend the weldment.

As shown, the small, unwelded end of the joint caused a crack to spread halfway along the weld as soon as it started to propagate. It only bent about 60 degrees before the crack formed.

Magnesium has very low elongation and poor formability, so the plate was tested to see how much it could bend before cracking. Heavy hammer blows were required to bend the plate. As shown, it bent with a large radius before a small crack was observed across the full width of the plate.

The cracked weld was fully broken and the surface examined. The weld was sound with no porosity evident. The fracture surface had a ductile appearance. This was an autogenous weld (no filler added) so the weld properties were expected to be less than the cold-rolled plate. It appears the weld had significantly less strength and little or no bending occurred in the base plate. The rapid crack propagation indicates possibly poor toughness, but the appearance of the broken surface did not indicate that was the case.

Magnesium has a large amount of spring-back when bent. As mentioned, the vice used to load the specimen required a high force to bend the welded joint. When the crack started, a high force was exerted on the test plate, which was probably the reason for the rapid crack progression. All of the deformation occurred in the weld, and there was no significant bending of the base material. It appears that the higher strength of the base plate and the high load needed to bend the weldment caused the initial unwelded section to form a crack in the weld, which rapidly progressed along the joint.

Weld Defects

Many welds have defects or discontinuities. Some of these discontinuities are very small and have little effect on weld performance. Some defects occur below the weld surface while externally the weld may look satisfactory. Descriptions of the most common discontinuities follow.

Lack of Fusion

Lack-of-fusion defects, also called cold laps, occur when there is no fusion between the weld metal and the surfaces of the base plate, as shown in Figure 7.14. Poor welding technique is the most common cause of lack of fusion. Either the weld puddle is too large (travel speed too slow) and/or the weld metal has been permitted to roll in front of the arc.

Weld metal leading the arc is a common problem for short-circuit MIG welding. In order to prevent this type of defect, the arc must be kept on the leading edge of the puddle. This prevents it from becoming too large and cushioning the arc.

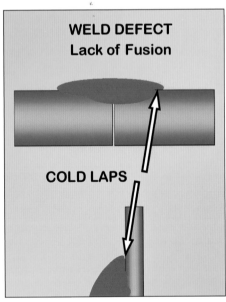

Fig. 7.14. Lack-of-fusion defects are also called cold laps. They occur when there is no fusion between the weld metal and the surfaces of the base plate. The most common cause of lack of fusion is a poor welding technique.

A very wide weld joint is another cause of lack of fusion. If the arc is directed down the center of the joint, the molten weld metal may only flow and cast against the side walls without melting them. The heat of the arc must be used to melt the base plate. To accomplish this you can make the joint narrower or direct the arc toward the side wall of the base plate by using a weaving technique.

Lack of fusion can also occur in the form of a rolled-over bead crown. This is generally caused by a very low travel speed or attempting to make too large a weld in a single pass. It may also be caused by too low a welding voltage, resulting in poor wetting of the weld bead. When welding aluminum, a common cause of this type of defect is the presence of aluminum oxide. This oxide has a very high

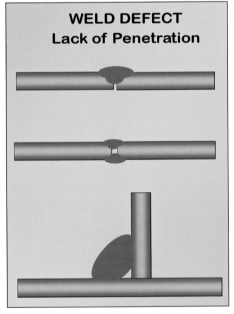

Fig. 7.15. Lack-of-penetration defects are created when the weld bead does not penetrate the entire thickness of the base plate (top), two opposing weld beads do not interlock (middle), or a weld bead does not penetrate the toe of a fillet weld but only bridges across it (bottom).

melting point and is insoluble in molten aluminum. If this oxide is present on the surfaces to be welded, fusion with the weld metal is hampered.

Lack of Penetration

Three lack-of-penetration defects are: a.) the weld bead does not penetrate the entire thickness of the base plate; b.) two opposing weld beads do not interlock; and c.) the weld bead does not penetrate the toe of a fillet weld but only bridges across it.

Welding current has the greatest effect on penetration. Welding current lower than necessary often causes incomplete penetration; simply increasing the amperage resolves the problem. Too-slow travel speed and an incorrect torch angle are other common causes. Both allow the molten weld metal to roll in front of the arc, acting as a cushion to prevent penetration. To prevent this and other defects, the arc must be kept on the leading edge of the weld puddle.

Undercutting

Undercutting is a common weld defect that appears as a groove in the parent metal along the edges of the weld. This type of defect is mostly caused by improper welding parameters, particularly travel speed and arc voltage. When the travel speed is too fast, the weld bead is peaked because of its quick solidification. Surface tension pulls the molten metal from the edges of the weld puddle back into the center. The undercut groove contains melted base material that has been drawn into the weld and not allowed to wet back properly.

Decreasing the travel speed gradually reduces the size of the undercut and eventually eliminates it. When only small or intermittent undercuts are present, raising the arc voltage

or using a leading torch angle also reduces undercut. In both cases the weld bead becomes flatter and wetting improves.

On the other hand, if the arc voltage is raised too high, undercutting may again appear because the arc melts more base metal than the weld puddle can fill. This is particularly true in MIG spray-arc welding. Excessive welding current can also cause undercutting. The arc force and penetration are so great that the molten metal is pushed to the rear of the weld puddle. The outermost areas of the base material are melted but solidify before the weld metal can fill the void. Puddle turbulence

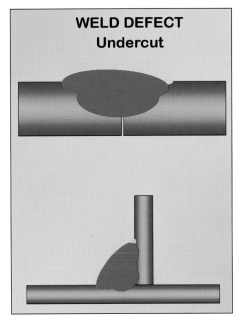

Fig. 7.16. Undercutting is a common weld defect that appears as a groove in the parent metal along the edges of the weld. This type of defect is mostly caused by improper welding parameters, particularly excess travel speed and arc voltage. Excessive welding current can also cause undercutting. The arc force and penetration are high and the molten metal is pulled by surface tension to the center rear of the weld puddle.

and surface tension pull the metal to the rear preventing proper wetting.

Porosity

Porosity is caused by gas bubbles that form and are trapped in the solidified weld bead. As seen in Figure 7.17, these pores may vary in size and may be distributed randomly in the weld root area. It is also possible the porosity is only found at the weld center, which is the last to solidify. Pores can occur either under or on the weld surface or both.

The most common causes of porosity are: a.) atmosphere contamination; b.) excessively rusty or scaled workpiece surfaces; and c.) inadequate deoxidizing alloys in the wire and the presence of foreign matter such as oil, water, paint, etc.

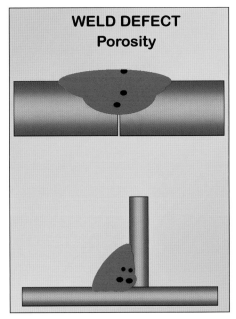

Fig. 7.17. Porosity is caused by gas pores that form and are trapped in the solidified weld bead. These pores may vary in size and may be distributed randomly in the weld root area. It is also possible that porosity is only found at the weld center, which is the last to solidify. Pores can occur under or on the weld surface, or both.

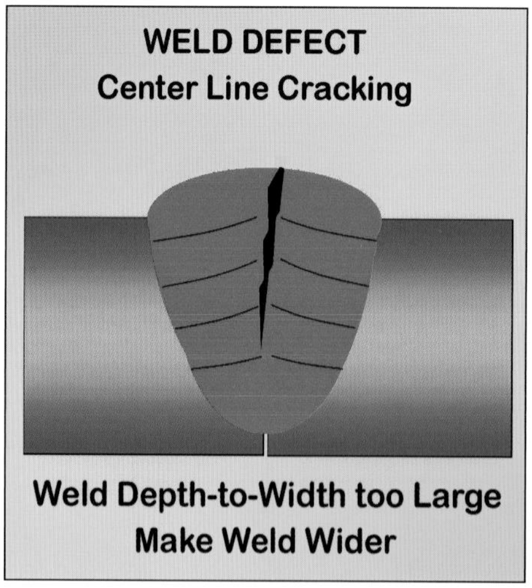

Atmospheric contamination can be caused by: a.) inadequate shielding gas flow; b.) excessive shielding gas flow that causes air to be pulled into the gas stream; c.) clogged gas nozzle or damaged gas supply system (leaking hoses, fittings, etc.); and d.) excessive wind in the welding area. High wind speeds or a draft can blow away the gas shield. If drafts exceed 4 to 5 mph when MIG welding, a windbreak must be used.

The atmospheric gases that are primarily responsible for porosity in steel are nitrogen, excessive oxygen, and hydrogen from water vapor, etc. When welding steel, some oxygen can be tolerated without porosity. However, oxygen in the atmosphere can cause severe problems when welding aluminum because of its rapid oxide formation. The shielding gas supply system should be inspected at regular intervals to ensure freedom from leakage. A gas leak-out means moisture-laden air is leaking back into the gas line through the same hole

Excessive moisture in the air can cause porosity in steel and (particularly) aluminum, so more care should

WELD DEFECT
Center Line Cracking

Weld Depth-to-Width too Large Make Weld Wider

Fig. 7.18. In addition to hydrogen cracking, longitudinal centerline cracking can occur. These are called hot cracks. Hot cracks occur while the weld bead is in the process of solidifying. They may result from the use of an incorrect electrode type when welding aluminum or stainless-steel alloys. An excessively deep weld compared to its width can cause a hot crack.

be exercised in humid climates. Rust and mill scale on the workpiece surface are obvious sources of oxygen as well as entrapped moisture.

Inadequate wire deoxidation can cause porosity. For example, ER70S-6 and ER70S-7 MIG wires have more deoxidizers than the more common ER70S-3 wire. The oxygen that enters the weld pool combines with carbon to form carbon gas porosity. Deoxidizers in the wire combine with the oxygen to form small particles of harmless manganese and silicon oxides. Some of these float to the sur-

face and form a glass-like slag. Excessive wire-drawing lubricant on the welding wire is a hydrocarbon. These hydrocarbons are sources of hydrogen, which is particularly a problem when welding steel and aluminum.

In addition to hydrogen cracking, longitudinal centerline cracking can occur. These are usually referred to as hot cracks, which occur while the weld bead is in the process of solidifying. Hot cracks may result from the use of an incorrect wire electrode when welding aluminum or stainless-steel alloys. The chemistry of the base plate can also promote this defect (an example would be steel having a high sulfur content or excess copper). Any combination of

WELD DEFECTS AND SOLUTIONS	
WELD POROSITY	A. OIL, RUST, SCALE, ETC. ON PLATE B. WIRE /ROD– MAY NEED WIRE HIGHER Mn AND Si C. SHIELDING PROBLEM; WIND, CLOGGED OR SMALL NOZZLE, DAMAGED GAS HOSE, EXCESSIVE GAS FLOW, LEAK IN GAS FITTINGS, ETC. D. FAILURE TO REMOVE SLAG BETWEEN WELD PASSES E. WELDING OVER SLAG FROM PREVIOUS STICK WELD TACK
LACK OF PENETRATION	A. WELD JOINT TOO NARROW B. WELDING CURRENT TOO LOW; C. TOO MUCH ELECTRODE STICKOUT D. WELD PUDDLE ROLLING IN FRONT OF THE ARC
LACK OF FUSION	A. WELDING VOLTAGE AND/OR CURRENT TOO LOW B. WRONG MIG POLARITY, SHOULD BE DCRP C. TRAVEL SPEED TOO LOW D. WELDING OVER CONVEX BEAD E. TORCH OSCILLATION TOO WIDE OR TOO NARROW F. EXCESSIVE OXIDE ON PLATE

WELD DEFECTS AND SOLUTIONS CONT.	
UNDERCUT	A. TRAVEL SPEED TOO HIGH B. WELDING VOLTAGE TOO HIGH C. EXCESSIVE WELDING CURRENT D. INSUFFICIENT DWELL AT EDGE OF WELD BEAD
CRACKING	A. INCORRECT WIRE/ROD CHEMISTRY B. WELD BEAD TOO SMALL C. POOR QUALITY QF MATERIAL BEING WELDED
UNSTABLE MIG ARC	A. CHECK GAS SHIELDING B. CHECK MIG WIRE FEED SYSTEM
POOR MIG WELD STARTS OR WIRE STUBBING	A. WELDING VOLTAGE TOO LOW B. INDUCTANCE OR SLOPE TOO HIGH C. WIRE EXTENSION TOO LONG D. CLEAN GLASS OR OXIDE FROM PLATE
EXCESSIVE MIG SPATTER	A. USE Ar-CO_2 OR Ar-O_2 INSTEAD OF CO_2 B. DECREASE PERCENTAGE OF He C. ARC VOLTAGE TOO LOW D. INCREASE INDUCTANCE AND/OR SLOPE

WELD DEFECTS AND SOLUTIONS CONT.	
BURN-THROUGH	A. WELDING CURRENT TOO HIGH B. TRAVEL SPEED TOO LOW C. DECREASE WIDTH OF ROOT OPENING
CONVEX BEAD	A. WELDING VOLTAGE AND/OR CURRENT TOO LOW B. EXCESSIVE MIG WIRE EXTENSION C. INCREASE MIG INDUCTANCE D. WRONG MIG POLARITY, SHOULD BE DCRP E. WELD JOINT TOO NARROW
CENTER BEAD CRACKING	A. WELD DEEPER THAN WIDE; INCREASE WELD WIDTH B. MOVMENT OF WELD WHILE SOLIDIFYING C. LOW MELTING MATERIAL SUCH AS SULFER OR EXCESS COPPER PRESENT. LOW MELTING POINT MATERIAL MOVES TO WELD CENTER AND AS WELD SOLIDIFIES LIQUID IS LEFT CREATING A CRACK.

Fig. 7.19. (Figure adapted from information produced by ESAB)

the joint design, welding conditions, and welding techniques that results in a weld bead with an excessively concave surface can promote cracking. One form of this defect is called a crater crack if it only forms at the end of the weld. This can often be solved by reversing the travel direction or pausing for a second or two before shutting off the arc.

Another cause of porosity is when the weld bead is narrow and deep (see Figure 7.18). Solidification occurs from the edges toward the centerline of the weld deposit. Impurities such as sulfur are pushed in that direction. This can create a material with a lower melting point, which remains liquid when the weld fully solidifies.

Figure 7.19 is a series of charts adapted from information supplied by ESAB, which provide a quick reference when troubleshooting weld defects.

A few comments are worth noting. A common weld defect is visible and internal porosity. In some instances, such as fillet welding, the porosity may be below the surface and is only visible by breaking a test section to look for these gas holes.

MIG short-arc welding can produce lack-of-fusion defects, and while the weld may look solid, it has not melted into parts of the base plate.

Excess spatter when MIG welding is not a defect because the weld quality is unaffected, but it causes a number of other problems. First, the spatter should be removed, and if the product is to be painted, it must be removed. Fine spatter is difficult to see, but if painted over, it may detach and create a rust spot. It also clogs the MIG gun nozzle and may lodge in the gas diffuser, disrupting the smooth laminar gas flow essential to good

shielding. The shielding gas mixture is often the cause of excess spatter. The use of 100-percent carbon dioxide shielding is less expensive to buy, but the excess spatter generated often make it false economy.

Gas mixtures of 8- to 25-percent carbon dioxide with the rest argon, produce the lowest spatter. The lower percentages of carbon dioxide produce less spatter but are somewhat colder. A mixture that increases the weld puddle fluidity is one containing 8-percent carbon dioxide, 2-percent oxygen, and the balance argon (see Chapter 6).

When welding with MIG short-circuiting, if the power source has adjustable slope or inductance, or different-level slope or inductance taps, it can dramatically reduce spatter by limiting the short-circuit current.

Project: Arc Straightening for Roll Bar Installation

A roll bar with cross brace is needed for our pro/street rod to anchor the shoulder harnesses. It must be removable to make it possible to finish the interior and build the sound wall, which replaced the back seat.

A fixture is fabricated to align all of the pieces for welding the simple fabricated roll bar. Two 90-degree bends and straight sections of 1¾-inch-diameter, .093-inch wall steel tubes were purchased from S&W Race Cars. The bottom ends of the roll-bar hoop need to slip into pipe stubs welded to the chassis and which protruded through the floor. Allowance was made for weld shrinkage—but not enough.

Figure 7.21 shows the chassis with two of the pipe stubs welded

to the top corners of the frame just before the narrowed rear frame section. However, even after allowing for some weld shrinkage, the distance between the vertical roll-bar tubes was 3/16 inch too short (as shown in Figure 7.22).

Weld distortion is used to increase the distance and correct the measurement. Flame straightening is sometimes used to bend metal without using mechanical force; however, this is an art and requires heating an area until it is red hot and rapidly cooling to provide local shrinkage. If the bend is excessive or in the wrong direction, it is difficult to correct. Arc straightening is easier to apply because small movements occur with each weld pass.

Small MIG weld beads were

placed on the outside edge of one vertical member (Figure 7.23). As the weld cools to room temperature, from the 2,500 degree F melting point, it attempts to shrink .018 inch for every inch of weld. However, the rigid assembly prevents it from moving that amount and creates some internal stress.

The full 3/16-inch movement was achieved after the fourth weld pass. One reason so few welds were needed was the location of the weld deposits relative to the long distance to the end of the vertical tubes. They were placed near the cross brace that had already been fillet welded to the vertical members. A small amount of shrinkage where they were located created much more movement at the tube end where it was needed.

The weld beads were ground flush with the tube when finished, which works fine for the mild-steel roll-bar tubing. If this were chrome-moly, preheating the area would reduce the chance of any hard, brittle material (martensite) forming in the HAZ.

Figure 7.25 shows the finished bar installed when testing the fit-ment. It fits through the door with the seats removed. It was removed when the interior was installed, leaving room to finish all surfaces. It serves the function of supporting the shoulder harnesses and provides moderate protection.

When fabricating custom rear differentials, Troy Spackman of Leg-acy Motors uses TIG welds to obtain perfect positioning of the two rear-axle end bearings. He uses a long, straight shaft to check alignment. When needed, small TIG welds are placed on the axle housing to create a slight bend in the direction of the weld. A great way to use weld distor-tion to advantage!

1 Fig. 7.20. Fabricate a fixture to align all of the pieces for welding a simple roll bar in a street rod. The bar has to be removable and include a cross brace to anchor the shoulder harnesses. The bottom ends of the roll bar need to slip into pipe stubs welded to the chassis that protrude through the floor. Be sure to make allowance made for weld shrinkage.

2 Fig. 7.21. Weld two of the pipe stubs to top corners of the frame just before the narrowed rear section. In this example, the distance between the vertical roll-bar tubes was 3/16 inch too short, even after allowing for some distortion due to weld shrinkage.

3 Fig. 7.22. To increase the distance without direct force, you can use what is normally considered weld distortion to correct the measurement. Flame straighten-ing is sometimes used to bend metal without using force; this is an art and requires heating an area until it is red hot and rapidly cooling to provide local shrinkage. Arc straightening is easier and more predictable to apply.

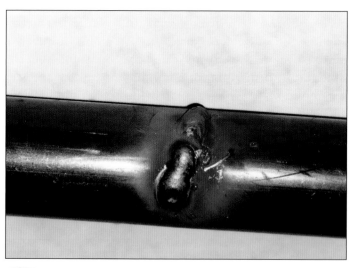

4 Fig. 7.23. Place small MIG weld beads on the outside edge of one vertical member. As the weld cools from 2,500 degrees F to room temperature, it bends that side of the bar slightly. Measure the movement after each small weld. Four weld passes are all that is needed to achieve the full 3/16-inch movement required.

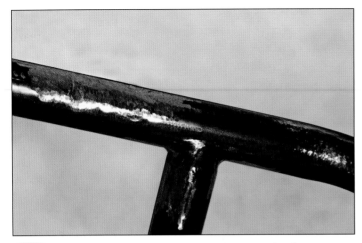

5 *Fig. 7.24. Grind the weld beads smooth when finished. Arc straightening works fine for this mild-steel roll-bar tubing. If this were chrome-moly, preheating the area would reduce the chance of any hard, brittle martensite forming in the HAZ.*

6 *Fig. 7.25. Test the finished roll bar for fitment. In this example, it fits through the door with the seats removed. Slip it in and bolt it to the pipe sections welded to the frame that protrude through the fiberglass. It serves the function of supporting the shoulder harnesses and provides moderate protection.*

7 *Fig. 7.26. Use the same technique as when fabricating custom rear differentials. To obtain perfect positioning of the two rear-axle end bearings, use a long, straight shaft to check alignment. When needed, place small TIG welds on the axle housing to create a slight bend. It bends in the direction of the weld.*

FURTHER READING

Irving, Bob. "Block Buster Events in Welding's Long History." *AWS Welding Journal* June 1999.

Jackson, C. E. "The Science of Arc Welding." *AWS Welding Journal* (Three Parts) April, May, June 1960.

Jackson, C. E. and A. E. Shrubsall. "Control of Penetration and Melting Ratio with Welding Technique." *AWS Welding Journal* April 1953.

———. "Energy Distribution in Electric Welding." *AWS Welding Journal* October 1959.

Lesnewich, Al. "Controlling the Melting Rate and Metal Transfer in Gas-Shielded Metal-Arc Welding. Part I." *AWS Welding Journal* August 1958.

Muller, Greene, and Rothschild. "Characteristics of Inert Gas Shielded Metal Arcs." *AWS Welding Journal* August 1952.

Uttrachi, G. D. "GMAW (MIG) Shielding Gas Flow Control Systems." *AWS Welding Journal* April 2007. (Also see NetWelding.com)

———. "NASCAR Race Team Demands Quality Welds." *AWS Welding Journal*, April 2003.

———. "Innovations in Cutting and Welding." 1997 National Steel Construction Conference (NSCC); Chicago, Illinois, May 1997.

———. "What Do Robots Need for Welding Equipment?" *Welding Design & Fabrication* June 1983.

Uttrachi, G. D. and D. Meyer "Basics of Semiautomatic (MIG) Welding." *AWS Welding Journal* August 1993.

U.S. Surgeon General, "Health of Arc Welders in Steel Ship Construction." Public Health Bulletin No. 298, 1947.

Wilson, Claussen and Jackson. "The Effect of I^2R Heating on Electrode Melting Rate." *AWS Welding Journal* January 1956.

SOURCE GUIDE

Affiliated Machinery
John Bray
3706 Alice St.
Pearland, TX 77581
affiliatedmachinery.com

Bruce Walters
4112 Byrnes Blvd.
Florence, SC 2950

ESAB Welding & Cutting Products
Bob Bitzky
411 South Ebenezer Rd.
Florence, SC 29501
esabna.com

Harvey Racing Engines
Jim Harvey 1
1 Toll Ln.
Levittown, NY 11756
hre.com

Legacy Innovations
Troy Spackman
1807 Andrew St.
West York, PA 17404
legacyinnovations.com

Rock Valley Antique Auto Parts
Dale Mathison
104 IL Hwy. 72
Stillman Valley, IL 61084
rockvalleyantiqueautoparts.com

S&W Race Cars
Jill Canuso
11 Mennonite Church Rd.
Spring City, PA 19475
swracecars.com

Stones's Welding
Randy Stone
3749 Half Moon Rd.
Pamplico, SC 29583

Total Cost Involved Engineering
Ben Bryce
1416 Brooks St.
Ontario, CA 81762
totalcostinvolved.com

Year One
Pat Staton
1001 Cherry Dr.
Braselton, GA 30517
yearone.com